Ariel Mesquita
Leonardo Bandeira
Claudia Martins

Rhizobacteria as bioinoculants

Ariel Mesquita
Leonardo Bandeira
Claudia Martins

Rhizobacteria as bioinoculants

The role of actinobacteria and rhizobacteria in promoting
plant growth

ScienciaScripts

Imprint

Any brand names and product names mentioned in this book are subject to trademark, brand or patent protection and are trademarks or registered trademarks of their respective holders. The use of brand names, product names, common names, trade names, product descriptions etc. even without a particular marking in this work is in no way to be construed to mean that such names may be regarded as unrestricted in respect of trademark and brand protection legislation and could thus be used by anyone.

Cover image: www.ingimage.com

This book is a translation from the original published under ISBN 978-620-6-76038-2.

Publisher:
Sciencia Scripts
is a trademark of
Dodo Books Indian Ocean Ltd. and OmniScriptum S.R.L publishing group

120 High Road, East Finchley, London, N2 9ED, United Kingdom
Str. Armeneasca 28/1, office 1, Chisinau MD-2012, Republic of Moldova, Europe
Printed at: see last page
ISBN: 978-620-7-66058-2

RHIZOBACTERIA AS BIOINOCULANTS: THE ROLE OF ACTINOBACTERIA AND RHIZOBIA IN PROMOTING PLANT GROWTH

SUMMARY

Actinobacteria, Gram-positive microorganisms found in various environments, are known for producing secondary metabolites and exoenzymes with biotechnological potential. Rhizobacteria, present in symbiosis with legumes, transform atmospheric nitrogen into ammonia, facilitating plant growth. This study co-inoculated strains of both bacterial groups isolated from the northeastern semi-arid region to investigate the influence of actinobacteria on rhizobial growth. Ten strains of actinobacteria and seven of rhizobacteria were selected, morphologically characterized and submitted to co-inoculation in culture media. The growth of the rhizobacteria was evaluated, showing a positive interaction, suggesting that the exoenzymes of the actinobacteria facilitate the metabolism of complex substrates by the rhizobacteria. This promising interaction could lead to the development of new bioinoculants to promote plant growth.

1 INTRODUCTION

The cowpea is the oldest cultivar known to man and has been planted for over four thousand years (OSIPITAN *et al.*, 2021). This bean is the source of much of the protein eaten in several African countries due to its low cost and growth in dry, nutrient-poor soils. This explains why this plant was the fourth largest source of dried grains in the world among legumes from 2008 to 2017 (JI *et al.*, 2019). Belonging to the Fabacea family, the cowpea, or *Vigna unguiculata*, grows mostly in the dry tropical soils of Latin America, Africa and South Asia (BOUKAR *et al.*, 2018).

This legume, like many others, has its roots associated with *rhizobia*, mainly from the genera *Bradyrhizobium* and *Rhizobium* (GIRIJA *et al.*, 2018). Rhizobia are bacteria known for forming symbiotic nodules on the roots (and, more rarely, the stems) of legumes and fixing atmospheric nitrogen in forms that can be assimilated by plants. This nutrient is one of the essential elements for maintaining life and is present in amino acids, nucleic acids, ATP and NAD in all living cells, as well as being an important component of chlorophyll (LINDSTRÖM; MOUSAVI, 2019). Nitrogen supplementation in crops is expensive and brings with it major pollution problems, such as eutrophication of aquifers and the release of greenhouse gases (CLÚA *et al.*, 2018).

Inoculation of rhizobia, which are natural bioinoculants, is an alternative to the use of these chemical fertilizers due to their ability to naturally bioavail nitrogen to plants through a process called biological nitrogen fixation (BNF). BNF is responsible for providing between 50 and 70 million tons of bioavailable nitrogen per year worldwide (SCHULTE *et al.*, 2021). Since 2020, Brazil has been the world's leading soybean producer, and inoculation of this legume with elite strains of *Bradyrhizobium* meets the plant's need for nitrogen and generates savings of US\$15 billion per year that would otherwise be spent on nitrogen fertilization (RODRIGUES *et al.*, 2020).

A lot is said about rhizobia to emphasize their symbiosis with plants and nitrogen fixation, causing a very important point to go unnoticed: the rhizospheric environment is extremely hostile for the microorganisms present there, and to be able to successfully establish symbiosis the rhizobia must first survive competition with the local microbiota (POOLE *et al.*, 2018).

With a relative abundance of approximately 22% and reaching up to 62% in desertified soils, the Actinobacteria phylum is one of the most abundant bacterial phyla in the soil and, consequently, interacts directly with rhizobia in the rhizosphere. These bacteria are able to increase moisture absorption in the soil and stimulate microbial growth (ARAUJO *et al.*, 2020).

Despite this stimulus, actinobacteria are producers of a wide variety of antimicrobial compounds. In fact, around two-thirds of known antibiotics are produced by actinobacteria, mainly by the genus *Streptomyces* (TAKAHASHI; NAKASHIMA, 2018). Therefore, the compatibility between rhizobia and actinobacteria is essential for the establishment of the symbiosis between the rhizobium and the plant and, consequently, for the proper growth and development of the plant. Based on these definitions, we can assume that the nitrogen-fixing capacity of rhizobia combined with the variety of secondary metabolites produced by actinobacteria can result in a promising bioinoculant for plant growth and development.

Bioinoculants are essential microorganisms for combating biotic and abiotic stresses in plants. Also called "plant growth-promoting rhizobacteria" (PGPR), these microorganisms are capable of carrying out activities such as hydrolysis of exudates in order to avoid osmotic stress, and the production of deaminases, indoleacetic acid, siderophores, phosphate-solubilizing enzymes and microbiocidal/biostatic enzymes (ENEBE; BABALOLA, 2018).

According to Dineshkumar and collaborators (2017), biofertilizers are "products containing living or dormant microorganisms (bacteria, actinobacteria, fungi, algae), alone or in combination, that help fix atmospheric nitrogen or solubilize soil nutrients". PGPR also benefit the plant through the production and release of growth-promoting substances, capable of increasing the plant's productivity. Some bioinoculants can present biofertilizing activity combined with antimicrobial activity, such as some species of *Pseudomonas* and *Bacillus* that stimulate plant growth while antagonizing pathogens and stimulating plant defenses (OROZCO-MOSQUEDA *et al.*, 2021).

For a microbial consortium to be able to stimulate plant growth, it is assumed that these microorganisms must first and foremost be compatible with each other. For this reason, it is worth carrying out *in vitro* co-inoculations before moving on to *in vivo* trials *in* association with plants. These steps are important both for excluding pairs of antagonistic microorganisms and for selecting pairs that show metabolic cooperation. The *in vitro* interactions between actinobacteria and rhizobia were studied in greater depth in this work, with emphasis on the potential of this consortium for the development of a bioinoculant capable of stimulating plant growth.

<u>**2 LITERATURE REVIEW**</u>

This section presents some important topics for the understanding and theoretical basis of the work. A literature review was carried out, presenting the state of the art with regard to actinobacteria, rhizobia, their interactions in the soil and their potential as bioinoculants.

2.1 Bacterial ecological interactions

Living beings generally interact when they have to share the same space, and it's no different with microorganisms. Throughout the growth of a colony, bacteria can specialize in different tasks, dividing and optimizing the work. The constant exchange of different metabolites between cells is the result of the colony's greater metabolic efficiency, which increases its chances of success (EVANS *et al.*, 2020).

Bacteria have up to 42% of their genes coding for characteristics involved in ecological relationships (PHELAN *et al.*, 2011). In nature, bacteria often compete for numerous limiting factors such as more favorable habitats, minerals and various nutrients. For this reason, these microorganisms have developed numerous strategies to enable growth and reproduction under these conditions, such as the secretion of toxins and antibiotics. The production of these metabolites gives an advantage to the growth of the bacteria that produced them, to the detriment of the rest of the local microbiota (D'SOUZA *et al.*, 2018).

In general, antagonistic relationships usually overlap with neutral relationships, which in turn overlap with positive relationships (LITTLE *et al.*, 2008). These interbacterial antagonistic relationships usually involve the secretion of small diffusible molecules (antibacterial secondary metabolites) and antibacterial protein toxins (antimicrobial peptides). When present in the same medium, it is quite common for bacteria to compete for nutrients in order to grow (KLEIN *et al.*, 2020). However, more complex interactions can occur, where the product of one strain's metabolism can be used for the growth of another. This positive ecological relationship is called "*cross-feeding*", metabolic cooperation or facilitation (SMITH *et al.*, 2019).

By transferring molecules via diffusion, bacteria are subject to the loss or degradation of these molecules or even consumption by third parties. For this reason, some bacteria have evolved to possess certain methods of delivering the desired molecules directly to the recipient. Methods such as outer membrane vesicles, channels, nanotubes, pili, or even membrane fusion with cytoplasmic exchange can be used to enable this exchange of metabolites more precisely (PANDE *et al.*, 2015). These methods, however, require the bacteria to maintain close physical contact. As this proximity is not always possible, it is necessary to adopt other methods so that the bacteria can cooperate.

Smith *et al.* (2019) dedicated a good part of their review on bacterial *cross-feeding to* classifying and describing various forms of metabolic cooperation between bacteria. In short, the authors define four general classes: metabolite *cross-feeding,* substrate cross-feeding, mutual cross-feeding and enhanced *cross-feeding.* The first case refers to when one bacterium takes advantage of the by-products of the metabolism of another, which feeds on complex carbon sources. Substrate cross-feeding occurs when a bacterium releases extracellular enzymes to degrade a certain substrate. The products of this degradation can be used by both the exoenzyme-producing bacterium and the opportunistic one. Mutual *cross-feeding* is nothing more than one of the previous examples, but when both bacteria feed and are fed by the other. Finally, increased cross-feeding is a type of mutual *cross-feeding,* where bacterium A intensifies the metabolic flow of synthesis of the product used by bacterium B, receiving by-products of B's metabolism in return.

2.2 *Cross-feeding* on the ground

Soil is a heterogeneous environment where it is common for biotic and abiotic factors to interact and influence each other. For example, the dynamics of nutrients such as carbon and nitrogen, as well as climate, have a direct effect on the diversity and ecology of soil microorganisms (MANDAL *et al.*, 2020). The abundance and diversity of soil bacteria and fungi is such that their metabolic activities have direct impacts on biogeochemical cycles at the biosphere level (SHARMA *et al.,* 2020).

There is a great diversity and complexity of carbohydrates in the soil. As there is a high metabolic cost to produce and excrete enzymes to hydrolyze these carbon sources, it is sometimes necessary for different species of microorganisms to cooperate to metabolize them. Given the variety of sugars in the soil, cooperative systems are common (LARSBRINK; MCKEE, 2020). There are also one-way routes, where "cheating" microorganisms take advantage of the product of reactions catalyzed by exoenzymes produced by other microorganisms, since these small molecules are able to diffuse away from the enzyme producer (VAN TATENHOVE-PEL *et al.*, 2021).

One example of metabolic cooperation in soil is the formation of multi-species biofilms. The heterogeneity of soil makes various structures available for the formation of biofilms, among which carbon-rich surfaces such as roots, fungal hyphae and decomposing organic matter are the most used by bacteria. Biofilms in soil are nothing more than an aggregate of different species of microorganisms that produce a matrix of extracellular polymeric substances (EPS). The proximity between microorganisms provided by the biofilm favors the exchange of metabolites, but this structure is more famous for protecting bacteria against environmental stresses, predation, dehydration and antibiotics, as well as improving the availability of nutrients and oxygen (CAI *et al.*, 2019).

From an evolutionary point of view, the sessile lifestyle provided by biofilms guarantees the various species of microorganisms that form it advantages over free-living microorganisms. Better growth during periods of dehydration and more opportunities for horizontal gene transfer are some examples. Although the positive points of biofilms are well known and studied, some characteristics such as the ecological and biological determinants for their formation, their influence on microbial metabolic activity and the structure of the communities formed are still unknown (WU *et al.*, 2019).

Rhizobacteria are also capable of forming biofilms, *with* the genera *Acetobacter, Alcaligenes, Bacillus, Pseudomonas, Rhizobium, Rhodococcus, Serratia* and *Streptomyces* being some examples of bacteria widely found in the rhizosphere with this ability. In general, rhizobacterial biofilms are the same as normal biofilms, but they have some ecological differences. Rhizobacteria naturally coexist in a consortium, interacting with each other, but they also influence the metabolism of plants. Multi-species rhizobacterial biofilms colonize roots and promote plant growth, as well as preventing attacks by pathogens, thus being of fundamental importance for sustainable agriculture (NAYAK *et al.*, 2020).

2.3 Actinobacteria and their biotechnological potential

Present mostly in soil, but also in oceans, extreme environments, plant tissues, animal excreta, algae and lichens (SINGH; DUBEY, 2018), the *Actinobacteria* phylum is one of the oldest in the Bacteria kingdom and played a vital role in colonizing the terrestrial environment (LAW *et al.*, 2020). These microorganisms are very diverse among themselves and among other bacteria (LEWIN *et al.*, 2016), which has led them to be called *Actinomycetes* for a long time due to their morphological similarity to filamentous fungi. The common characteristics between these two classes of microorganisms meant that, for a long time, actinobacteria were considered a transition between bacteria and fungi. However, with the advancement of technologies, it was observed that the morphology of "actinomycetes" was much more similar to that of Prokaryotes than to Eukaryotes, such as thin cells, with a chromosome stored in a nucleoid, a peptidoglycan cell wall and susceptibility to antibacterials (BARKA *et al.*, 2015).

Unlike bacteria in general, actinobacteria are capable of forming aerial and reverse (or substrate) mycelium and reproducing by sporulation (by fragmentation and segmentation or conidia formation) (ANANDAN *et al.*, 2016). The main forms of phenotypic differentiation between these microorganisms are the presence or absence of aerial and reverse mycelium, color of the mycelium, diffusion of pigments in the medium and the structure and appearance of their spore chains (BARKA *et al.*, 2015). Actinobacteria are Gram-positive and generally have high levels of guanine and cytosine (G + C) in their single chromosome, up to 70% (LEWIN *et al.*, 2016). Actinobacteria often have long linear plasmids, and some genera such as *Streptomyces, Actinomyces, Amycolatopsis, Actinoplanes, Streptoverticillium* and *Micromonospora* also have a linear chromosome (VENTURA *et al.*, 2007).

Actinobacteria have a relative abundance of 22% in soil, which is one of their main habitats. Desertified soils can have an abundance of up to 62% of these microorganisms (ARAUJO *et al.*, 2020), where they play vital roles such as stabilizing clay particles and organic matter, relieving biotic and abiotic stress on plants, fixing nitrogen, solubilizing phosphorus sources, decomposing plant residues (SOLANS *et al.*, 2021), hydrolyze complex and recalcitrant polymers (such as lignocellulose, keratin and chitin) and solubilize insect cuticles, crustacean shells and the cell wall of fungi and plants

(LACOMBE-HARVEY, 2018).

In hostile environments such as the semi-arid region, where it is common to see soils in the process of desertification, the survival of plants depends to a large extent on the community of microorganisms associated with them, either in the rhizosphere or in symbiosis. In this water-poor soil, microorganisms are essential for modifying the soil structure in order to optimize biological activity, for example by retaining water (SOLANS *et al.*, 2021). The characteristic smell of "wet soil" also comes from actinobacteria, which produce a terpene called geosmin, which is what gives this odor (SALWAN; SHARMA, 2020).

Some actinobacteria even act as endophytes, mainly of medicinal plants and in tropical forests. The roots of the host plant release exudates that directly impact the microbiota present in the rhizosphere (TANVIR *et al.*, 2018). When in symbiosis with plants (most commonly in the roots), actinobacteria act mainly by producing phytohormones or other growth factors, increasing resistance to biotic and abiotic stresses, insects, pests and pathogens in exchange for nutrients and shelter in the host plant (SINGH *et al.*, 2018). Actinobacteria are also capable of suppressing competitors by synthesizing antibiotics (BAO *et al.*, 2021). This diverse production of secondary metabolites currently justifies why around 45% of the 22,500 known compounds with biological activity are extracted from actinobacteria. In addition, these microorganisms produce 80% of known antibiotics, especially the genera *Streptomyces* and *Micromonospora* (MAJIDZADEH *et al.*, 2021).

The biotechnological value of the *Actinobacteria* phylum is undeniable. These microorganisms are famous for synthesizing and excreting secondary metabolites of high industrial interest, such as immunosuppressants, anticancer compounds, antivirals, antifungals, antiparasitics, anthelmintics and, as already mentioned, antibacterials (GOODFELLOW *et al.*, 2018; MAJIDZADEH *et al.*, 2021; VAN BERGEIJK, 2020). Actinobacteria also synthesize extracellular enzymes such as amylases, cellulases and xylanases (SATRIA *et al.*, 2020; FATMAWATI *et al.*, 2018).

Amylases are enzymes with potential applications in various fields, such as medical, textile and ethanol production, as well as in the treatment of fruits such as bananas, mangoes and citrus fruits, washing of bioreactors in the food industry, starch scarification and fermentation and distillation industries (AL-AGAMY *et al.*, 2021).

Cellulases are enzymes capable of hydrolyzing cellulose into glucose due to their exoglucanase, endoglucanase and beta-glucosidase subunits. This ability of cellulase makes it relevant in industrial processes such as textiles, food, animal feed improvement, detergent production, pulp and paper industry, extraction of green tea components and bioethanol production from lignocellulosic biomass (ISLAM; ROY, 2019; BHATI *et al.*, 2020). Lignocellulose, the world's largest source of plant biomass, is mostly made up of cellulose (35-50%) and hemicellulose (20-35%). Most hemicellulose is made up of xylan, which is the second most abundant component after cellulose (PUTRI; SETIAWAN, 2019).

Xylanase is responsible for converting xylan into xylose, thus degrading hemicellulose, an essential component of plant cell walls and making these nutrients available to microorganisms present in plant sources (ÇIÇEKLER, 2022). The degradation of lignocellulose from plant waste is the main industrial application of xylanases, where they are used in the production of biofuels, reducing the viscosity of juices and feed and in the paper industry, as a wood pulp bleaching agent for the production of better quality paper (KAUSHAL *et al.*, 2021).

The genus that stands out most in the production of these exoenzymes of industrial interest is *Streptomyces* (RACHMANIA *et al.*, 2020; KUMAR *et al.*, 2020; MIHAJLOVSKI *et al.*, 2020). This genus is the most abundant among the actinobacteria, and is also where the antibiotics streptomycin, gentamicin, rifamycin, chloramphenicol and erythromycin are extracted (AL-SHAIBANI *et al.*, 2021). The genus *Streptomyces* has 1134 species and 71 subspecies (PARTE *et al.*, 2020). Its importance is such that actinobacteria belonging to other genera are called "rare actinobacteria" (YIN *et al.*, 2018). This term, however, seems to be inappropriate, some authors such as Jose *et al.* (2019) suggest classifying these microorganisms as "*non-Streptomyces* actinobacteria", which better defines less abundant actinobacteria, but still highlights the importance of the dominant genus.

Streptomyces have a multicellular life cycle in which they undergo various morphological and physiological changes. After spore germination, the bacterium grows by extension of its extremities and forms a network of hyphae known as a reverse or substrate mycelium. The maturation of the colony leads to the differentiation of an aerial mycelium from cell division and the generation of monoploid spores. It is at this stage of differentiation that the famous secondary metabolites are produced, with the aim of allowing the individual to survive even in unfavorable environments (LAW *et al.*, 2019).

Because they contain several biosynthetic gene clusters (BGCs), actinobacteria have the metabolic flexibility needed to become promising targets for metabolic engineering to produce high value-added secondary metabolites (AL-SHAIBANI *et al.*, 2021). BGCs are genomic regions that contain three or more non-homologous genes encoding biosynthetic enzymes. The genes present in BGCs generally form a specific biosynthetic pathway and are usually expressed simultaneously (POLTURAK *et al.*, 2022).

2.4 Rhizobia

Another group of bacteria present in the soil, more specifically in the roots of legumes, are the rhizobacteria. The term "rhizobium" is an umbrella term to define bacteria from eight families, seven of which are α-Proteobacteria (Rhizobiaceae, Phyllobacteriaceae, Nitrobacteriaceae, Methylobacteriaceae, Bruccllaceae, Hyphomicrobiaceae and Xanthobacteriaceae), the most common, and one β Proteobacteria (Burkholderiaceae). Currently, 226 species of rhizobia are known, distributed in 19 genera: *Agrobacterium, Allorhizobium, Ensifer (Sinorhizobium), Neorhizobium, Pararhizobium, Rhizobium, Shinella, Aminobacter, Phyllobacterium, Mesorhizobium, Bradyrhizobium, Microviga, Methylobacterium, Brucella (Ochrobactrum), Devosia, Azorhizobium, Cupriavidus, Paraburkholderia* and *Trinickia* (HELENE *et al.*, 2022).

Rhizobia are Gram-negative aerobic bacteria that are mobile (have one to six peritrichial flagella), rod-shaped, non-spore forming, with mucous and convex colonies ranging from white to beige, and from shiny to opaque (KUYKENDALL *et al.*, 2015). In general, rhizobia are classified according to their growth time (fast or slow) and effect on the pH of the YMA (yeast extract agar-mannitol) culture medium. Fast-growing species usually acidify the pH of the medium, while slow-growing species do not usually alter the pH of this medium (SADOWSKI *et al.*, 1983)

The importance of these microorganisms is due to their ability to establish a symbiotic relationship with legume roots and induce the formation of nodules, where they will carry out biological nitrogen fixation (BNF). Although nitrogen is the most abundant gas in the atmosphere, it is not bioavailable to plants. In search of nitrogen, legumes release flavonoids through their roots, attracting rhizobia to the rhizosphere, where they secrete nodulation factors (YANG *et al.*, 2021). The rhizobia attach themselves to the roots, forming an infection pocket, and secrete cellulases to penetrate the roots. Thus, the plant will give rise to a nodule meristem, which will develop and attract the bacterial cells to its interior. Finally, the bacteria are trapped in organelle-like structures called symbiosomes, where they will differentiate into bacterioids and finish maturing into nodules. Within the microaerophilic environment of these mature nodules, the rhizobia are ready to reduce atmospheric nitrogen (N_2) into ammonia by means of nitrogenase enzymes (POOLE *et al.*, 2018).

The nitrogenase enzyme complex is where nitrogen fixation occurs in all known diazotrophic organisms. Rhizobia have *nif* genes in their genome, which are responsible for synthesizing the nitrogenase complex and various nitrogen fixation regulatory enzymes, which can also be encoded by *fix* genes (LINDSTRÖM; MOUSAVI, 2019). The nitrogenase complex is made up of two proteins: one, dinitrogenase reductase (ferroprotein), has an Fe cluster S_{44} , and the other, dinitrogenase (molybdenum-ferroprotein), has Fe clusters S_{87} and FeMo cofactors (MoFe S_{79} C-homocitrate). Therefore, nitrogenase synthesis and consequently nitrogen fixation are dependent on sulphur, and its presence in the soil is directly linked to the rhizobium-plant symbiotic relationship (SCHNEIDER *et al.*, 2018).

The final form of the nitrogen fixed by the rhizobium will depend on the legume with which it is associated. In temperate plants, ammonia is converted into glutamine and asparagine, and in tropical environments into ureides such as allantoin and allantoic acid. Nitrogen is made bioavailable to plants via the xylem, which in return provide carbon and energy sources in one of nature's most famous symbiotic relationships (BOSSE *et al.*, 2021). The diversity of final forms of nitrogen goes hand in hand with the diversity of legumes colonized by rhizobia, which consists of 750 plant genera (RAJKUMARI *et al.*, 2022).

Nodulation is mainly controlled by the plant, which can even prevent nodules colonized by rhizobia with low nitrogen-fixing efficiency from remaining. For this reason, some rhizobia can develop mechanisms and strategies to improve their ability to nodulate and fix nitrogen in a given plant (BASILE; LEPEK, 2021). BNF is the most efficient way of supplying nitrogen to plants. In addition to saving billions of dollars spent annually on industrialized nitrogen fertilizers, which are polluting and originate from fossil (non-renewable) sources, the use of rhizobia as nitrogen providers for plants is a much more accessible alternative for subsistence farmers (DICENZO *et al.*, 2018).

2.5 Bioinoculants

The Green Revolution consisted of a set of initiatives adopted worldwide to improve agricultural production, mainly in developing countries. This was achieved by improving irrigation, using large quantities of chemical nutrients and planting high-yielding cultivars. However, these strategies have caused serious environmental problems, such as a decrease in the biological and physicochemical health of the soil, loss of biodiversity, genetic erosion, ecological imbalance, decreased tolerance to plant stress, among others (MAITRA *et al.*, 2021). One strategy to reduce the use of synthetic agrochemicals and combat this soil deterioration would be to take advantage of the genetic and biological potential of cultivars and microorganisms associated with plants (BASU *et al.*, 2021).

Endophytic microorganisms are those that colonize plant tissues without causing disease and establish a beneficial relationship with their host, producing phytohormones, enzymes, antagonizing pathogens, stimulating phytoremediation and bioavailing nutrients (ALKAHTANI *et al.*, 2020; CHAUDHARY *et al.,* 2020). These beneficial rhizobacteria include rhizobia, some actinobacteria, mycorrhizal fungi and free-living bacteria (DUTTA; PODILE, 2010).

Bacteria that live in the rhizosphere and stimulate plant growth through one or more mechanisms, even in the presence of competitors, are called "plant growth-promoting rhizobacteria" (PGPR) and represent 2 to 5% of the rhizospheric microbiota (OO *et al.*, 2020). PGPR can be extracellular (ePGPR), living in the rhizosphere, in the rhizoplane, or in the spaces between the cells of the plant cortex. Some examples of ePGPR are bacteria of the genus *Agrobacterium, Arthrobacter, Bacillus, Burkholderia, Erwinia, Micrococcus, Pseudomonas* and *Serratia*. Intracellular PGPR (iPGPR) generally colonize specialized structures (nodules) on the roots and are usually from the genera *Allorhizobium, Bradyrhizobium, Mesorhizobium, Rhizobium,* among others (PRASAD *et al.*, 2019).

Actinobacteria are capable of acting as PGPR in both direct and indirect ways. The direct method includes the production of siderophores and phytochromes, phosphate solubilization and nitrogen fixation, while the indirect method includes the synthesis of lytic exoenzymes, antibiotics, volatile compounds and competing with pathogens for nutrients (HAYAT *et al.*, 2020). Actinobacteria from the genera *Streptomyces, Thermobifida, Frankia, Nocardia, Kitasatospora, Micromonospora, Actinomadura, Streptosporangium, Actinoplanes* have been reported as PGPR with the most varied applications (FRANCO-CORREA; CHAVARRO-ANZOLA, 2016).

All the microorganisms mentioned so far in this article can be considered examples of bioinoculants. By definition, bioinoculants are microorganisms that stimulate plant growth by colonizing its root system. In general, these microorganisms act to absorb nutrients and protect against diseases, but they can even be applied to the bioremediation of pollutants. The effect of bioinoculants on plant health is also related to the mechanism of induced systemic resistance (ISR), which is activated by the JA/ET (jasmonic acid/ethylene) and salicylic acid pathways (CHAUDHARY; SHUKLA, 2019). It is for all these benefits that it becomes evident that PGPR-based formulations are a promising alternative to minimize the use of synthetic fertilizers and agrochemicals in agriculture, which in the long term can cause soil acidification and reduced nutrient absorption by the roots (LAU *et al.*, 2020).

One negative aspect of using bioinoculants in agriculture comes from an ecological point of view, since PGPR inoculation can affect the native microbiota. These interactions can be positive or negative, and will depend on the physical and chemical characteristics of the soil and other abiotic conditions. Changes in the population of microorganisms in that area can affect the quality and fertility of the soil, which is why it is in the industry's interest to better understand the interactions between bioinoculants and native microorganisms in order to maximize their effectiveness (MANFREDINI *et al.*, 2021).

The semantics of the terms bioinoculant, biofertilizer and biopesticide vary greatly in the literature, often being taken as synonyms (PHATAK; KUMAR, 2016). However, bioinoculants are a general term for microorganisms that are beneficial to plants, and they can be classified according to their action as biofertilizers, if they improve the bioavailability of nutrients in the soil, and as biopesticides, if they antagonize phytopathogens. Table 1 shows some of the bioinoculants known in agriculture.

TABLE 1 - List of some known bioinoculants and their respective **applications**

Biofertilizers		
Function	**Type**	**Name**
Nitrogen fixers	Bacteria	*Rhizobium, Azotobacter, Azospirillum, Frankia*
Zinc solubilizers		*Bacillus* sp., *Pseudomonas* sp., *Enterobacter* sp., *Mycobacterium* sp.
Phosphate solubilizers		*Bacillus megaterium, Pseudomonas* sp., *Rhodococcus, Serratia, Micrococcus*
Phosphate solubilizers	Fungi	Arbuscular mycorrhizae, *Penicillium, Piriformospora indica*
Micronutrient mobilizers		Arbuscular mycorrhizae
Biopesticides		
Name	**Type**	**Targets**
Bacillus sp.	Bacteria	*Fusarium, Verticillium, Ascochyta, Alternaria, Xanthomonas, Erwinia*
Bacillus thuringiensis		Caterpillars, beetles, leafhoppers, insects
Serratia sp.		*Sclerotium*
Pseudomonas sp.		*Penicillium, Botrytis cinerea, Mucor, Pythium*
Gliocladium sp.	Fungi	*Penicillium, Aspergillus, Fusarium*
Trichoderma sp.		*Fusarium, Macrophomina, Ascochyta, Cercospora*
Verticillium lecanii		Thrips, whiteflies, aphids, mealybugs
Nuclear polyhedrosis virus (NPV)	Virus	Caterpillars, earthworms, moths, corn borer

Source: KUMAR, 2018

The use of rhizobia as biofertilizers for legumes is a well-established technology, due to the ability of these bacteria to reduce the need for nitrogen fertilization as a result of the increase in nitrogen absorption by plants promoted by SNF (LADAN *et al.*, 2022). Symbiotic relationships between rhizobia and plants have already been well studied, from *quorum-sensing* for nodule formation to the directed evolution of rhizobia in symbiosis towards mutualism with the plant. It is now known that plant growth stimulated by PGPR generally does not occur through direct nitrogen fixation, but through the production of phytohormones (MIA; SHAMSUDDIN, 2010).

Legumes have been known for centuries to have a positive effect on soil fertility, and only later was it discovered that this effect was actually the result of the biofertilizing action of rhizobia (PALAI *et al.*, 2021). The progress of studies in this area led to the filing of the first patent for a *Rhizobium-based* biofertilizer in 1896, which was named Nitragin (NOBBE; HILTNER, 1896). The use of mixed rhizobial bioinoculants with other PGPRs are considered the "ultimate inoculants", due to their potential for developing new commercial products. However, there is still some hesitation on the part of farmers to use bioinoculants. This is because commercial formulations are often established without much scientific rigor, leading to products with little impact on soil fertility and crop yields. Despite this, there is a consensus in the scientific community that biofertilizers are cheap, sustainable and effective products that remedy the negative impacts caused by chemical fertilization (ATIENO *et al.*, 2020).

Several examples of successful co-inoculation of rhizobia and other PGPRs can be found in the literature. *Bradyrhizobium*, for example, show positive results in plant growth and development when co-inoculated with *Pseudomonas oryzihabitans, Pseudomonas putida, Bacillus megaterium, Bacillus pumillus, mycorrhizae (Glomus clarum, Glomus mosseae, Gigaspora margarita)* (KUMAWAT *et al.*, 2022; MILJAKOVIĆ *et al.*, 2022; JABBOROVA *et al.*, 2021; SHETEIWY *et al.*, 2021), among others.

Actinobacteria have a positive effect on the nodulation and growth of legumes, such as the co-inoculation of soybeans with *Bradyrhizobium japonicum* and *Streptomyces* sp. and *Nocardia* sp., and alfalfa with *Sinorhizobium meliloti* and *Micromonospora* spp. or *Frankia*, which stimulates nodulation even in soils with high nitrogen levels, a condition that usually inhibits nodulation (SAIDI *et al.*, 2021). The search for bioinoculants based on actinobacteria and rhizobia, more specifically between the genera *Streptomyces* and *Bradyrhizobium*, is well documented in the literature (HTWE *et al.*, 2019; SOE; YAMAKAWA, 2013; HTWE *et al.*, 2018; HTWE; YAMAKAWA, 2016). Given this, it is clear that these two genera stand out when it comes to prospecting for new PGPR with bioinoculant potential. The variety of plants and their respective edaphic microbiota observed around the world makes it necessary to use the most suitable bioinoculants for each location and climatic conditions. For example, microorganisms isolated from semi-arid areas are more suitable for biotechnological application in cultivars used in this climate, so it is necessary to study the co-inoculation of microorganisms at a local level.

<u>**3 OBJECTIVES**</u>

The general and specific objectives of the research are presented below.

3.1 General objective

Coinoculate strains of actinobacteria and rhizobia in order to prospect a new bioinoculant for use in agriculture, by evaluating the capacity for metabolic cooperation between these two different bacterial phyla isolated from soils in Brazilian semi-arid zones.

3.2 Specific objectives

a) Choose isolates of actinobacteria with significantly different high cellulolytic, amylolytic and xylanolytic activity;

b) Selecting rhizobial strains without cellulolytic, amylolytic and xylanolytic activity;

c) Characterize the macromorphology of actinobacteria in terms of the color of the aerial and reverse mycelia, colony shape and pigment production;

d) Characterize the micromorphology of actinobacteria in terms of the shape of the spore chains;

e) Infer the genus of actinobacteria from the structure of the hyphae and spores;

f) Coinoculating actinobacteria and rhizobia *in vitro*;

g) To choose the most compatible strains of actinobacteria and rhizobia, with a view to future *in vivo* tests.

4 MATERIAL AND METHODS

In order to achieve the established objectives, methodologies were adopted for obtaining the samples, characterizing and co-inoculating the microorganisms and statistical analysis. These strategies are detailed in this section.

4.1 Choice of strains

The microorganisms used in this study were obtained from the culture collection of the Environmental Microbiology Laboratory (LAMAB) at the Federal University of Ceará. Previous studies by our research group have evaluated the amylolytic, cellulolytic and xylanolytic activity of actinobacteria and rhizobacteria from the culture collection (unpublished results). To choose the strains to be used, the enzymatic indices of 313 strains from this culture collection were subjected to an analysis of normality and multivariate ANOVA variance (MANOVA) using SPSS software (IBM Corp. Released 2011). To verify the hypotheses of the statistical tests used, the enzymatic indices of the actinobacteria were subjected to the Chi-square test. The normality of the data was analyzed using the Kolmogorov-Smirnov and Shapiro-Wilk tests. The homogeneity of variances was analyzed using Levene's test. To check for differences between the models, the enzyme indices were subjected to a Student's t-test for unpaired data. The means of the enzymes of each strain were compared using an analysis of variance using the Tukey HSD test. All statistical tests were carried out at a significance level of 0.05.

For this work, 10 strains of actinobacteria with statistically distinct enzymatic activity (A108, A109, A125, A136, A139, A143, A144, A145, A146, A148) were chosen from a universe of 313 strains, and 7 rhizobacteria without enzymatic activity for the enzymes in question (L1, L4, L9, L13, L15, L24, L27) from a universe of 150 strains. These rhizobial strains were previously identified by sequencing the 16s rRNA molecule by Silva (2020) (Table 2).

TABLE 2 - Rhizobial strains from the Brazilian semi-arid region belonging to the culture collection of the Environmental Microbiology Laboratory (LAMAB) and selected for the co-inoculation tests.

Cepa	Species
L1	*Bradyrhizobium elkanii*
L4	*Bradyrhizobium elkanii*
L9	*Rhizobium tropici*
L13	*Bradyrhizobium kavangense*
L15	*Bradyrhizobium japonicum*
L24	*Bradyrhizobium yuanmingense*
L27	*Bradyrhizobium iriomotense*

Source: Silva (2020).

The soil samples that originated the actinobacteria used in this work were collected in accordance with the research project accepted by CNPq/ICMBio/FAP's Notice No. 18/2017, process 421350/2017.2[1] . A108, A109 and A125 were obtained from soils in preserved areas of the Aiuaba Ecological Station - CE (A108 and A109 at 6° 07'S, 40° 2'W; A125 at 6° 75'S, 40° 2'W). A136 and A139 come from soils in preserved areas of the Ubajara National Park - CE (3° 5'S, 40° 5'W). A143, A144, A145 and A146 were isolated from soils in preserved areas of the Sete Cidades National Park - PI (8° 2'S, 42° 4'W). A148 was obtained from soils in secondary areas also in the Sete Cidades National Park - PI (8° 2'S, 42° 4'W).

4.1.1 Repicking the chosen strains

The strains chosen were replicated from the collection for use. The actinobacteria strains were inoculated by depletion in CDA (casein-dextrose-agar)

---

[1] Responsible for biological material: MSc. Leonardo Lima Bandeira

medium and CD (casein-dextrose) broth and incubated in a B.O.D. incubator at 28°C for 7 days. The rhizobial strains were inoculated by depletion in YMA (yeast extract-mannitol-agar) medium and YM (yeast extract-mannitol) broth and incubated under the same conditions and for the same length of time. The composition of the culture media is described in Table 3.

TABLE 3 - Details of the components and ideal pH of the culture media used in the research

Culture medium	Components	pH
CDA	0.5 g.L^{-1} of K$_2$HPO$_4$, 0.2 g.L^{-1} of MgSO$_4$, 2 g.L^{-1} of glucose, 0.01 g.L^{-1} of FeSO$_4$, 0.2 g.L^{-1} of casein, 15 g.L^{-1} of agar, 2 mL.L^{-1} of nystatin 100,000 UI.mL^{-1}	6,6
CD	0.5 g.L^{-1} of K$_2$HPO$_4$, 0.2 g.L^{-1} of MgSO$_4$, 2 g.L^{-1} of glucose, 0.01 g.L^{-1} of FeSO$_4$, 0.2 g.L^{-1} of casein	6,6
YMA	10 g.L^{-1} mannitol, 0.5 g.L^{-1} K$_2$HPO$_4$, 0.2 g.L^{-1} MgSO$_4$, 0.1 g.L^{-1} NaCl, 0.5 g.L^{-1} yeast extract, 5 mL.L^{-1} bromothymol blue 0.5% in KOH 0.2N, 15 g.L^{-1} agar, 2 mL.L^{-1} nystatin 100,000 IU.mL^{-1}	6,8
YM	10 g.L^{-1} mannitol, 0.5 g.L^{-1} K$_2$HPO$_4$, 0.2 g.L^{-1} MgSO$_4$, 0.1 g.L^{-1} NaCl, 0.5 g.L^{-1} yeast extract, 5 mL.L^{-1} bromothymol blue 0.5% in KOH 0.2N	6,8
CMC	5 g.L^{-1} of carboxymethyl cellulose, 0.5 g.L^{-1} of MgSO$_4$, 0.5 g.L^{-1} of KCl, 3 g.L^{-1} of NaNO$_3$, 0.01 g.L^{-1} of FeSO$_4$, 1 g.L^{-1} of K$_2$HPO$_4$, 15 g.L^{-1} of agar, 2 mL.L^{-1} of nystatin 100,000 UI.mL^{-1}	6
AM	10 g.L^{-1} of peptone, 3 g.L^{-1} of meat extract, 5 g.L^{-1} of NaCl, 2 g.L^{-1} of starch, 15 g.L^{-1} of agar. 2 mL.L^{-1} of nystatin 100,000 UI.mL^{-1}	6,8 ± 0,3
XY	1 g.L^{-1} of xylan obtained from wood, 0.5 g.L^{-1} of MgSO$_4$, 1 g.L^{-1} of yeast extract, 0.5 g.L^{-1} of NaNO$_3$, 0.01 g.L^{-1} of FeSO$_4$, 1 g.L^{-1} of K$_2$HPO$_4$, 15 g.L^{-1} of agar, 2 mL.L^{-1} of nystatin 100,000 UI.mL^{-1}	6,5

| ISP 2 | 4 g.L^{-1} yeast extract, 10 g.L^{-1} malt extract, 4 g.L^{-1} dextrose, 20 g.L^{-1} agar, 2 mL.L^{-1} nystatin 100,000 IU.mL^{-1} | 7,0 ± 0,2 |

Source: the author

4.2 Characterization of the strains

The actinobacteria strains were characterized in terms of the nature of their cell wall by Gram staining. Macro and micromorphological characterization was also carried out, the protocols for which are detailed below.

4.2.1 Gram stain

In order to observe the filamentous morphology, as well as to assess the purity of the inocula and to obtain preliminary confirmation that the bacterium is in fact an actinobacterium, a Gram stain (Êxodo Científica®, Sumaré-SP) was carried out following the manufacturer's instructions. Using a sterile inoculation loop, a smear was made in saline solution from an isolated colony previously grown in CDA medium. This smear was subjected to 1 minute in gentian violet, 1 minute in lugol, washed with acetone alcohol and then 30 seconds in Gram's fuchsin. Finally, the slide was observed using a Leica optical microscope® DM750 at 1000x magnification.

4.2.2 Macromorphological cultural characterization

The strains underwent a cultural characterization of their macromorphology. The actinobacteria strains were inoculated by *spots* in CDA medium (Table 3), from which the colors of the aerial and reverse mycelium of the colonies were observed (which were then classified according to the RAL color chart) (SANTOS *et al.*, 2019a), the appearance of the colonies (SILVA *et al.*, 2019b), and the production of diffusible pigments (RAMOS *et al.*, 2015).

4.2.3 Micromorphological characterization

To identify the morphology of the spore chains of the actinobacteria, a micromorphological characterization (microcultivation) was carried out according to Santos *et al.* (2019b) with modifications. A cube of CDA medium with a side of approximately 1 cm was transferred to a slide sheltered inside a sterile Petri dish. The strains were inoculated onto the sides of this cube, which was then covered with a coverslip. Two pieces of absorbent cotton moistened with sterile distilled water were placed inside the plate to preserve humidity. The plates were incubated in a B.O.D. oven at 28°C for 10 days, then the coverslips containing the fingerprints of the hyphae were transferred to a new sterile slide and stained with Amann's lactophenol (Êxodo Científica® , Sumaré-SP). The spore chains were observed using a Leica® DM750 optical microscope at 1000x magnification and compared with the structures described by Dornelas *et al.* (2017) to identify the microstructure. The genus of the actinobacteria was inferred according to The Society For Actinomycetes Japan (1997).

4.3 *In* vitro facilitation

An *in vitro* co-inoculation was carried out to investigate the capacity for metabolic cooperation (facilitation) between the strains of actinobacteria and rhizobia according to the methodology of Silva *et al.* (2019a) modified by Mesquita *et al.* (2022). For this, three culture media were used, each containing a single carbon source: carboxymethylcellulose (CMC), starch (AM) and xylan (XY), as shown in Table 3. Firstly, the actinobacteria were inoculated by *spot in* duplicate onto the aforementioned culture media and incubated for 7 days in a B.O.D. incubator at 28°C. At the end of the 7 days, the plates were assessed for the presence of contamination and adequate growth. Contaminated plates or those with no growth were discarded and repeated.

The rhizobia were then purified for inoculation. 1 mL of each rhizobium in YM broth (Table 3) was transferred to a sterile microtube and centrifuged in a Marconi MA 1800 centrifuge at 9261 x g (10300 RPM) for 10 minutes. The supernatant was discarded and the *pellet* resuspended in 1 mL of distilled water and homogenized in a Phoenix AP56 vortex mixer. This process was repeated twice, in triplicate, until the purified rhizobia suspended in distilled water were obtained.

Immediately after purifying the rhizobacteria, the actual co-inoculation took place, where 10 µL of the purified rhizobacteria were inoculated next to the actinobacteria *spots*. Growth was reassessed after a further 7 days of incubation in a B.O.D. incubator at 28°C. The growth of the rhizobium colonies characterized a positive result.

The compatibility index (CI) was calculated from the ratio between the number of compatible pairs and the number of possible pairs. ICA would be the index for actinobacteria, and ICR for rhizobacteria. The antagonism index (AI) was calculated from the ratio between the number of antagonistic pairs and the number of possible pairs. IAA would be the index for actinobacteria, and IAR for rhizobacteria.

4.4 Selecting the most promising strains

Based on the compatibility and antagonism indices (CI and AI, respectively), statistical tests were carried out using SPSS software (IBM Corp. Released 2011) to determine the most promising strains for future *in vivo* trials. To verify the hypotheses of the statistical tests used, CI and IA were analyzed by a Chi-square test using a binary matrix, and then by a Kruskal-Wallis test for independent samples. A k-means test was carried out for CI and AI, from which the data was divided into 3 *clusters and* a dendrogram was plotted. The two actinobacteria with the greatest similarity to each other and the highest CI index were chosen for future *in vivo* trials. All statistical tests were carried out at a significance level of 0.05.

4.5 Scanning electron microscopy

Scanning electron microscopy (SEM) was carried out in order to observe the microstructure of the hyphae and spores in more detail. Only the strains with the greatest potential for *in vivo* testing were observed by SEM. The samples were prepared according to the methodology of Balagurunathan *et al.* (2020). Sterile aluminum *stubs* were inserted at an angle of approximately 45° into a plate with solid ISP 2 medium (Table 3) until approximately half of the *stub was* inside the medium. The same procedure was carried out with glass coverslips. The plates were then incubated for 24 hours in a B.O.D. incubator at 37°C to check for contamination. After this time, the actinobacteria were inoculated with a sterile loop along the line where the *stub* and the coverslips meet the medium. The plates were then incubated for around 10 days at 28°C in a B.O.D. incubator.

The coverslips were prepared for viewing via optical microscopy following the same methodology as the microcultivation. This step is used to observe the maturity of the spores, which must be fully developed before SEM can be carried out. When the spores reached a satisfactory level of maturity, the *stubs* were carefully removed from the medium and sent to the Analytical Center of the Federal University of Ceará, as they are part of the approved project "Morphology of actinobacteria strains from the northeastern semi-arid region". The samples were fixed and metallized at the Analytical Center, and electron microscopy was carried out using QUANTA™ 450 FEG equipment at 5,000, 10,000 and 70,000 times magnification.

5 RESULTS AND DISCUSSION

The results obtained through this research are presented and discussed in this section, highlighting the potential of co-inoculation of actinobacteria and rhizobia as bioinoculants.

5.1 Characterization of the strains

The strains used in this work were characterized in terms of their macro and micromorphology, and a Gram stain was carried out to analyze the purity of the inoculums. The results obtained are presented in this section.

5.1.1 Gram stain

Actinobacteria show significant morphological differentiation from other bacteria in this class, but the most common are filamentous (HAZARIKA; THAKUR, 2020). Most actinobacteria have a cell membrane with a thick layer of peptidoglycan, which characterizes them as Gram-positive (RAHLWES *et al.*, 2019). All the strains retained their crystal violet color (Figure 1), confirming that the inocula are pure and that the bacteria are indeed Gram-positive.

FIGURE 1 - Gram staining of actinobacteria strains under 1000x magnification in an optical microscope.

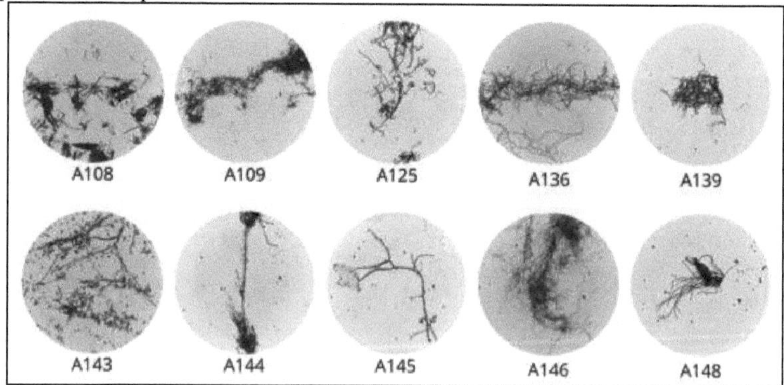

Source: the author.

The Gram stain method was developed in 1884 by Hans Christian Gram and innovated the field of microbiology by making it possible to observe stained bacteria via light microscopy (GRAM, 1884). In this method, which differentiates bacteria according to their cell wall, Gram-positive bacteria appear purple, while Gram-negative bacteria usually vary between pink and reddish (ROHDE, 2019). To this day, almost 140 years after the technique was developed, Gram staining is still considered the gold standard for classifying bacteria in terms of the physical properties of the cell wall (KWON *et al.*, 2019).

5.1.2 Macromorphological cultural characterization

Due to the diversity of colors in the same colony (Figure 2), the strains were characterized in terms of the colors of the edges and ends of each colony. The results are summarized in Table 1. As for the production of diffusible pigments in the medium, none of the strains showed this ability when grown in CDA medium. Graph 1 summarizes the results obtained in relation to the appearance of the colonies.

FIGURE 2 - Color diversity of actinobacteria aerial mycelium

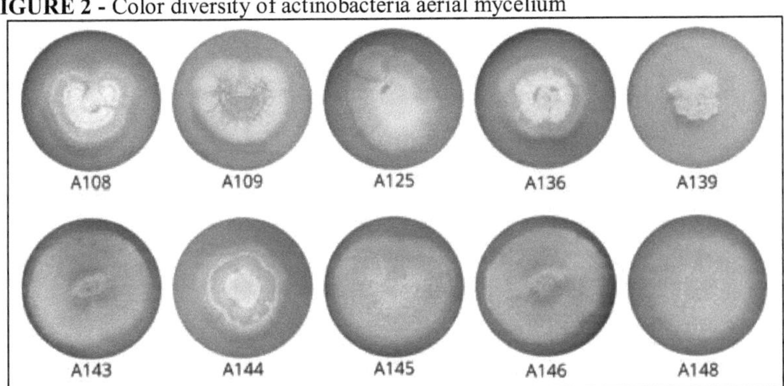

Source: the author.

TABLE 1 - Macromorphological cultural characteristics of actinobacteria strains

Strains	Aerial Mycelium		Reverse mycelium		Appearance of the colony
	Center	**Edge**	**Center**	**Edge**	

A108	RAL 9003 Signal white	RAL 7044 Silk grey	RAL 9010 Pure white	RAL 9018 Papyrus white	Concentric
A109	RAL 8001 Ochre brown	RAL 9003 Signal white	RAL 8023 Orange brown	RAL 9003 Signal white	Radial with grooves
A125	RAL 9001 Cream	RAL 9003 Signal white	RAL 1002 Sand yellow	RAL 9003 Signal white	Radial with grooves
A136	RAL 1002 Sand yellow	RAL 1000 Green beige	RAL 9010 Pure white	RAL 9003 Signal white	Concentric
A139	RAL 3012 Beige red		RAL 3022 Salmon pink		Umbonada
A143	RAL 8001 Ochre brown	RAL 9010 Pure white	RAL 8029 Pearl copper	RAL 9003 Signal white	Velvety
A144	RAL 7044 Silk grey	RAL 9002 Grey white	RAL 1002 Sand yellow	RAL 7038 Agate grey	Concentric
A145	RAL 1034 Pastel yellow	RAL 1013 Oyster white	RAL 7044 Silk grey	RAL 7035 Light grey	Concentric
A146	RAL 8001 Ochre brown	RAL 9010 Pure white	RAL 8029 Pearl copper	RAL 9003 Signal white	Velvety
A148	RAL 9002 Grey white		RAL 7047 Telegrey 4		Concentric

Source: the author.

GRAPH 1 - Appearance of actinobacteria colonies.

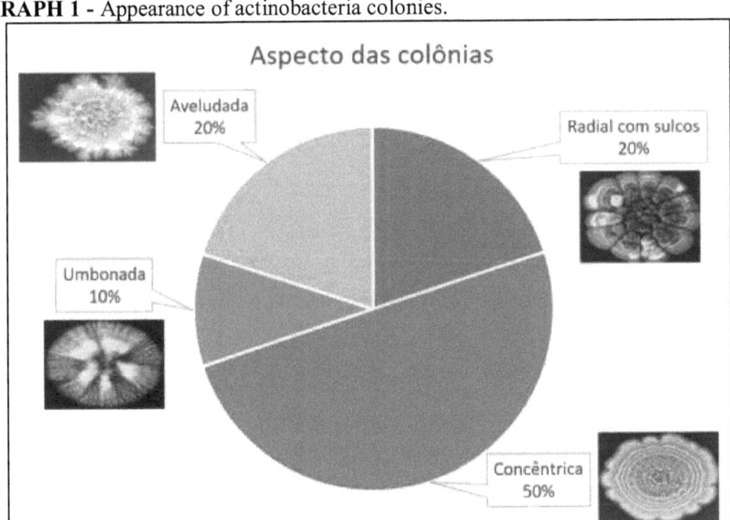

Source: SILVA *et al.*, 2019b; the author.

Actinobacteria are capable of producing extra- and intracellular pigments (e.g. in the mycelium and spores). When cultivated for the required time, these microorganisms form colonies with aerial and reverse mycelia that can present varied colors (AXENOV-GIBANOV *et al.*, 2016). These characteristics, however, do not only depend on the genetic potential of each strain, but are directly influenced by the incubation conditions and nutritional composition of the culture medium in which they are growing. The ISP2 culture medium is the most widely used for cultural characterization of actinobacteria (BEHIE *et al.*, 2017). Despite this, the characterization was carried out on CDA medium, which according to the results obtained by our research group is more suitable for the growth of actinobacteria from the Brazilian semi-arid region.

The pigments produced by actinobacteria can be water-soluble or fat-soluble. Water-soluble pigments color the medium with their respective color, while fat-soluble pigments color the colony itself. In addition, the color of these pigments can vary between white, brown, black, red, yellow and orange (HAZARIKA; THAKUR, 2020). From this, it is possible to state that the strains used are capable of producing a diversity of fat-soluble pigments when cultivated in CDA medium, with their predominantly white, brown and yellow colors matching what is described in the literature. The presence of grey pigments may be characteristic of actinobacteria isolated from the semi-arid region, such as the strains characterized by Medeiros *et al.* (2018), Lima *et al.* (2017) and Ramos *et al. (*2015).

5.1.3 Micromorphological characterization

One of the main characteristics used to study the taxonomy of actinobacteria at genus and species level is their micromorphology. With regard to spores, these can be formed in both the aerial and reverse mycelium, in single cells (*Micromonospora*) or large chains that can have hundreds of spores (*Streptomyces*). Some genera can form specialized vesicles (sporangia), which are basically spore pockets (*Frankia*) (BARKA *et al.*, 2015). Figure 3 summarizes the structures observed in the microcultivation, which are mainly spore chains.

FIGURE 3 - Spores of the actinobacteria strains

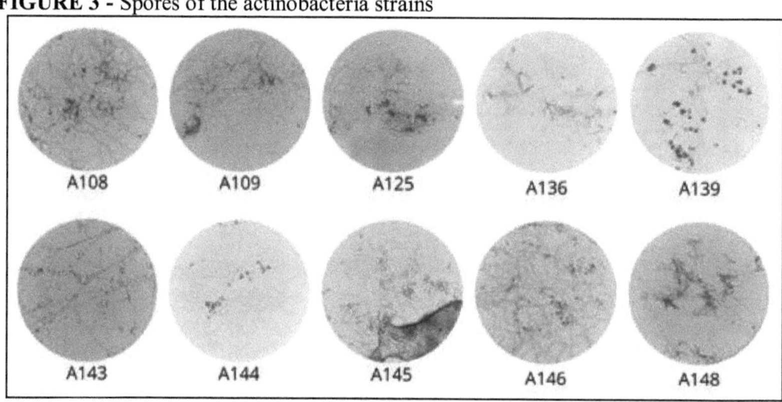

Source: the author.

From these structures, the type of spore chains was determined and the genus of each strain was inferred (Table 2). The diversity of spore chain types and genera is visually represented in Graphs 2 and 3, respectively.

 Micromorphological characterization of actinobacteria strains

Strains	Type of spore chains	Probable gender
A108	Spirals	*Streptomyces*
A109	Flexible	*Nocardia*
A125	Open spirals	*Nocardia*
A136	Spirals	*Streptomyces*
A139	Spirals	*Streptosporangium*
A143	Straight	*Micromonospora*
A144	Spirals	*Streptosporangium*
A145	Open spirals	*Streptomyces*
A146	Spirals	*Streptosporangium*
A148	Open spirals	*Streptomyces*

Source: the author.

GRAPH 2 - Genus diversity of actinobacteria strains

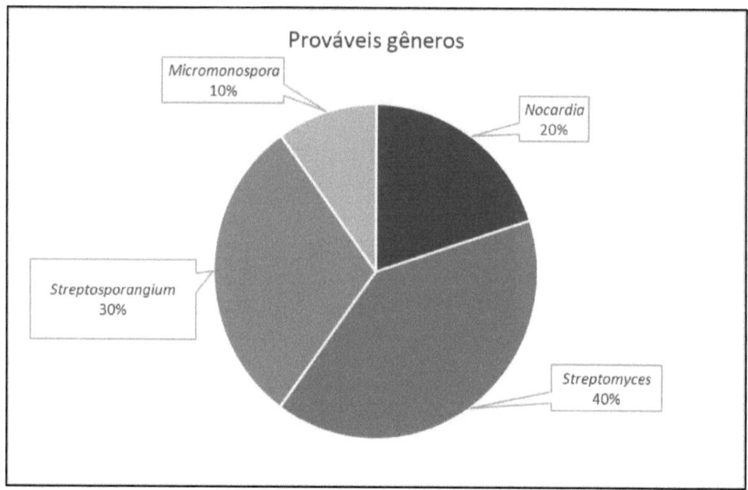

Source: the author.

GRAPH 3 - Diversity of actinobacteria spore chains

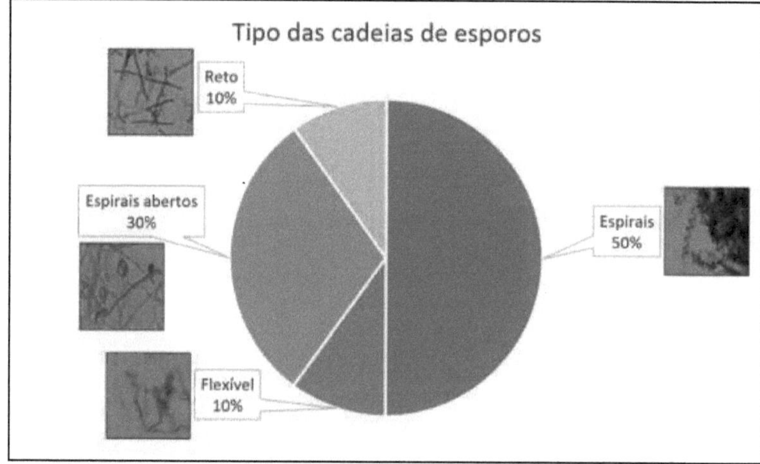

Source: DORNELAS *et al.* (2017); the author.

Spore chains can be divided morphologically according to the number of spores: bispores with two spores, oligospores with few spores and polyspores with several spores. The groups that produce only one spore per "chain" are called monospores and their main representatives are the genera *Micromonospora* (as in strain A143), *Thermomonospora, Saccharomonospora* and *Thermoactinomyces* (LI *et al.*, 2016).

The presence of sporangia was observed in strains A144 and A146, which were classified as *Streptosporangium*. These structures are responsible for developing and releasing spores and vary greatly in shape and size, and can be formed in both the aerial and reverse mycelium. *Streptosporangium* sporangia are usually spherical with a spiral, unbranched spore chain and, as with most spore-forming actinobacteria, planospores, although there are exceptions (HAMEDI; POORINMOHAMMAD, 2017; KURTBÖKE *et al.,* 2020).

The typical morphology of the genus *Nocardia* consists of filamentous and branched cells. Often, only bead-like cells are visible, with these structures organized along branched filaments (BROWN-ELLIOT *et al.,* 2015). These characteristics were observed in the microcultivation of strains A109 and A125, allowing them to be classified as belonging to the genus *Nocardia*.

The genus *Streptomyces* is known for its complex life cycle, from spore germination to the formation of vegetative and aerial mycelia, and new spore chains (sporulation) (PACIOS-MICHELENA *et al.*, 2021). Actinobacteria of this genus have polyspores (arthrospores) that form long chains of more than 50 spores. These spores are formed by the septation and disarticulation of pre-existing hyphae (LI *et al.*, 2016). According to Zhuang *et al.* (2020), a classic characteristic of bacteria of the genus *Streptomyces* is the spiral spore chains, as was observed in the microcultivation of strains A108, A136, A139, A145 and A148.

Of the 53 actinobacteria that Parihar *et al.* (2022) isolated from the Thar Desert in India, 34 (64%) were *Streptomyces*. Other studies involving the micromorphological characterization of actinobacteria from soils in the northeastern semi-arid region have observed different abundances to those observed in this study. Santos *et al.* (2019b), for example, found 37% abundance of *Streptomyces* and 13% of *Nocardia* in open areas, and 75% of *Streptomyces* and 25% of *Nocardia* in closed areas in the Médio Jaguaribe micro-region. Of the 14 strains of actinobacteria that Lima *et al.* (2017) isolated from the Aiuaba Ecological Station, 13 belonged to the genus *Streptomyces* and one to the genus *Saccharothrix*. Although other genera have been observed, it is clear that *Streptomyces* is the dominant genus of actinobacteria in semi-arid soils (HAMID *et al.*, 2020).

This study obtained 50% *Streptomyces*, 20% *Streptosporangium*, 20% *Nocardia* and 10% *Micromonospora*. This is mainly due to the extremely small size of the sample space used here. It is not possible to study microbial diversity accurately with just 10 strains, and this is not the focus of this study. The micromorphological characterization was carried out only to observe more characteristics of the strains and infer their genera according to the structure of their spores.

5.2 *In* vitro facilitation

Some pairs, such as actinobacterium A146 and rhizobium L4 in the medium with CMC (Figure 4) showed a result similar to an antagonistic relationship, where the actinobacterium produced a halo inhibiting the growth of the rhizobium. These results were tabulated as negative.

FIGURE 4 - Example of antagonistic interaction between rhizobium strain L4 (left) and actinobacterium strain A146 (right) in CMC medium.

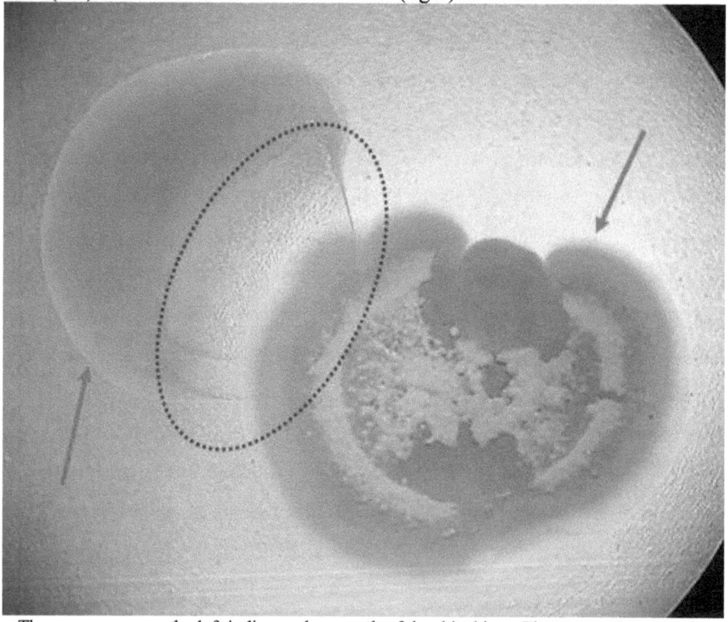

The green arrow on the left indicates the growth of the rhizobium. The orange arrow on the

right indicates the growth of the actinobacterium. The blue dotted line highlights the halo of
inhibition of rhizobium growth produced by the actinobacterium.
Source: the author.

In nature, bacteria often compete for numerous limiting factors such as more favorable habitats, minerals and various nutrients. For this reason, these microorganisms have developed numerous strategies to enable growth and reproduction under these conditions, such as secreting toxins and antibiotics, which give an advantage to the growth of the bacteria that produced them (D'SOUZA *et al.*, 2018) . In general, these antagonistic relationships usually overlap with neutral relationships, which in turn overlap with positive relationships such as *cross-feeding* (LITTLE *et al.,* 2008).

According to Klein *et al.* (2020), one of the modes of action of interbacterial antagonism is the secretion of small diffusible molecules (antibacterial secondary metabolites) and antibacterial protein toxins (antimicrobial peptides). Actinobacteria A136, A139, A145, A148 (*Streptomyces* sp.), A146 (*Streptosporangium sp.),* and A143 (*Micromonospora* sp.) showed antagonism towards at least one of the rhizobacteria tested.

Compounds extracted from actinobacteria are responsible for producing 90% of known antibiotics for commercial use. The genus *Streptomyces*, in addition to being the largest genus of actinobacteria, is also the most famous producer of compounds with antibiotic activity (DAS *et al.*, 2018; NOFIANI; RIZKY; BRILLIANTORO, 2021). However, rare actinobacteria such as those of the genera *Micromonospora* (GEORGE *et al.*, 2012) and *Streptosporangium* (VERMA *et al.,* 2008) also show antibacterial activity, which is in line with the results obtained in this work.

Figure 5 shows the positive interaction between actinobacterium strain A143 and rhizobium L27 in the xylan medium, while Figure 6 shows a negative result between actinobacterium strain A145 and rhizobium L27 in the starch medium.

FIGURE 5 - Positive interaction between rhizobium strain L27 (left) and actinobacterium strain A143 (right) in xylan medium. Example of a positive result.

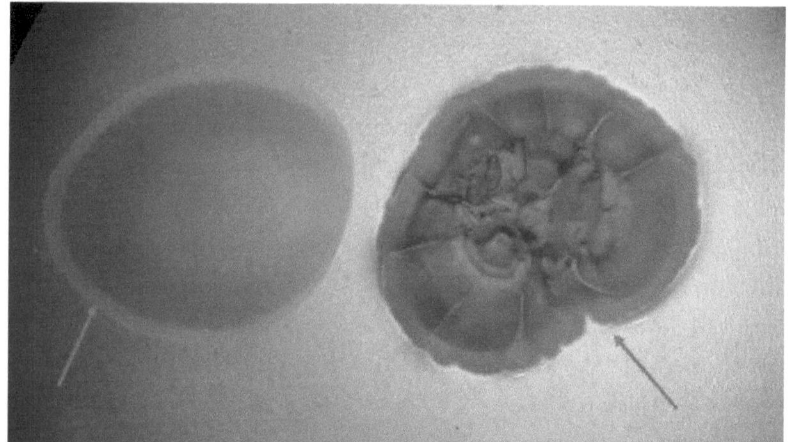

The green arrow on the left indicates the growth of the rhizobium. The orange arrow on the right indicates the growth of the actinobacterium.
Source: the author.

FIGURE 6 - Absence of interaction between rhizobium strain L27 (inoculated on the left side of the image, where there is no growth) and actinobacterium strain A145 (right) in the starch medium. Example of a negative result.

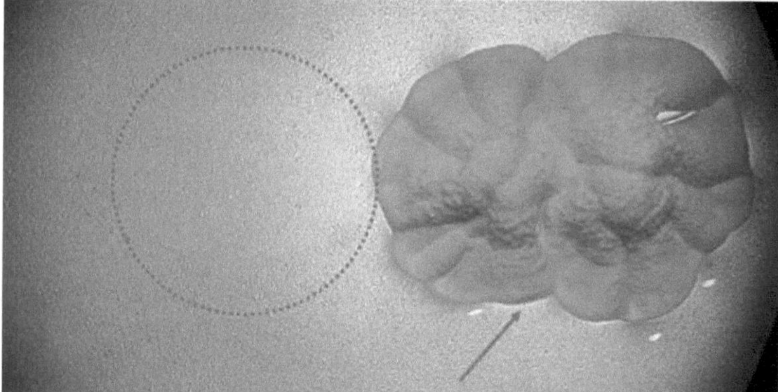

The green dotted line on the left indicates where the rhizobium was inoculated, highlighting the absence of growth. The orange arrow on the right indicates the growth of the actinobacterium. Source: the author.

In the medium with CMC, A144 and A148 tested positive with all the rhizobia tested (ICA = 1). A108, A139 and A145 showed an ICA of 0.857, presenting a positive result with 6 of the 7 rhizobacteria strains tested. The L1 strain showed growth with 9 of the 10 actinobacteria tested (ICR = 0.9) (Table 3).

TABLE 3 - *In vitro* facilitation between strains of actinobacteria and rhizobia from the Brazilian semi-arid region in CMC medium.

Rhizobium	Actinobacteria										ICR	IAR
	A108	A109	A125	A136	A139	A143	A144	A145	A146	A148		
L1	+	+	+	+	+	+	+	+	-	+	0,9	0
L4	+	-	+	A	+	+	+	+	A	+	0,7	0,2
L9	-	-	-	-	A	-	+	A	-	+	0,2	0,2
L13	+	-	-	-	+	-	+	+	-	+	0,5	0
L15	+	+	+	A	+	-	+	+	A	+	0,7	0,2
L24	+	-	-	+	+	-	+	+	+	+	0,7	0
L27	+	-	-	+	+	+	+	+	+	+	0,8	0
ICA	0,857	0,286	0,429	0,429	0,857	0,429	1	0,857	0,286	1		
IAA	0	0	0	0,286	0,143	0	0	0,143	0,286	0		

(+) presence of facilitation (positive result). (-) absence of facilitation (negative result). (A) antagonistic interaction, also considered a negative result.
Source: the author.

In the medium with xylan, A109, A139 and A145 tested positive with all the rhizobia tested (ICA = 1). Strains A125 and A143 had an ICA of 0.857, showing a positive result with 6 of the 7 rhizobium strains. Rhizobium strain L4 showed growth with 9 of the 10 actinobacteria tested (ICR = 0.9) (Table 4).

TABLE 4 - *In vitro* facilitation between strains of actinobacteria and rhizobia from the Brazilian semi-arid region in the medium with xylan.

Rhizobium	Actinobacteria										ICR	IAR
	A108	A109	A125	A136	A139	A143	A144	A145	A146	A148		
L1	+	+	+	A	+	+	+	+	A	+	0,8	0,2
L4	+	+	+	+	+	+	+	+	+	A	0,9	0,1
L9	-	+	-	-	+	+	+	+	-	-	0,5	0
L13	-	+	+	A	+	A	+	+	A	-	0,5	0,3
L15	-	+	+	A	+	+	-	+	A	A	0,5	0,3
L24	+	+	+	+	+	+	-	+	A	+	0,8	0,1
L27	+	+	+	+	+	+	-	+	A	A	0,7	0,2
ICA	0,571	1	0,857	0,429	1	0,857	0,571	1	0,143	0,286		
IAA	0	0	0	0,429	0	0,149	0	0	0,714	0,429		

(+) presence of facilitation (positive result). (-) absence of facilitation (negative result). (A) antagonistic interaction, also considered a negative result.
Source: the author.

The medium with starch proved to be a less favorable environment for rhizobial growth, since only nine results out of the seventy possible were positive, especially actinobacterium A148, with 3 positive results (ICA = 0.429) and rhizobium L1, with 4 positive results (ICR = 0.4) (Table 5).

TABLE 5 - *In vitro* facilitation between strains of actinobacteria and rhizobia from the Brazilian semi-arid region in the starch medium.

Rhizobium	Actinobacteria										ICR	IAR
	A108	A109	A125	A136	A139	A143	A144	A145	A146	A148		
L1	-	-	-	-	-	+	-	+	+	+	0,4	0
L4	-	-	-	-	+	-	-	-	-	+	0,2	0
L9	-	-	-	-	-	-	-	-	-	-	0	0
L13	-	-	-	-	-	-	-	-	-	-	0	0
L15	-	-	-	-	+	-	-	-	-	-	0,1	0
L24	-	-	-	-	-	-	-	-	-	-	0	0
L27	-	-	+	-	-	-	-	-	-	+	0,2	0
ICA	0	0	0,143	0	0,286	0,143	0	0,143	0,143	0,429		
IAA	0	0	0	0	0	0	0	0	0	0		

(+) presence of facilitation (positive result). (-) absence of facilitation (negative result).
Source: the author.

SNF is an energetically costly process and depends on cellulose-rich plant residues to obtain this energy. In addition, the rhizobium needs to break through the plant cell wall to be able to colonize the plant and nodulate. However, rhizobia, such as the seven used in this study, do not usually have a high cellulolytic activity and do not use cellulose as a carbon source (SILVA *et al.*, 2019a). Given the mixed composition of lignocellulose, the largest constituent of plant cell walls, which is rich in cellulose and hemicellulose (WU *et al.,* 2019), and the inability of these bacteria to degrade these carbohydrates, it is clear that rhizobia need help to colonize the roots. We propose the use of actinobacteria isolated from the same site as the rhizobacteria to carry out this research on *cross-feeding,* since studies using bacteria isolated from the same environment have a greater similarity to the natural community, as highlighted by the work of Stadie *et al.* (2013). Thus, this choice more faithfully represents the interactions that would take place *in situ.*

Metabolic cooperation interactions between different bacterial phyla are ubiquitous in nature and play critical roles in determining the structure and function of microbial communities (PONOMAROVA; PATIL, 2015). However, despite the many studies on the subject, the rules governing the establishment of these interactions are still not fully understood.

In their work on metabolic cooperation in bacterial species, Giri *et al.* (2021) identified that metabolic dissimilarity between the genotype of the donor and the recipient is the main determinant for the establishment of these cross-interactions between two strains of bacteria, and that the growth of the recipient depends on the production of amino acids by the donor. The *Rhizobiaceae* family, which is home to rhizobia, is made up of Gram-negative, rod-shaped bacteria. Their main characteristic is the formation of symbiotic nodules with legumes, where they perform biological nitrogen fixation (KUYKENDALL, 2015; WHEATLEY *et al.*, 2020). Actinobacteria are Gram-positive filamentous microorganisms capable of forming aerial and substrate mycelium and sporulating (JOSE *et al.*, 2021). The differentiation of this phylum from other bacterial phyla is so old that it is currently not possible to specify their most closely related groups (GOBALAKRISHNAN, 2022). The presence of facilitation between actinobacteria and rhizobia evidenced by this work corroborates the results of Giri *et al.* (2021), given the physiological, morphological and phylogenetic disparity between these organisms.

While these inferences can be corroborated by our work, where mainly the cocultures in media with carboxymethylcellulose and xylan showed positive results for metabolic cooperation, the results in starch media leave room for some questions. The main one is to try to understand how in a medium with a homopolysaccharide as a carbon source, with a relatively simpler chain than cellulose and xylan (ABE *et al.* 2022), predominantly negative results were observed for metabolic cooperation.

In this sense, the research by Zheng *et al.* (2021), which relates the presence and intensification of fluorosurfactants and amylase in soil bacteria, brings pertinent conclusions to the results of co-inoculation in starch medium. According to the authors, the presence of α-amylase strongly influences the adaptation strategies of the bacteria's metabolism, in addition to the production of metabolic signal factors (such as acetic acid), resulting in changes in physicochemical properties such as hydrophobicity and surface charge of the bacterial surface and consequently in microbial interaction. As all the actinobacteria strains in this study are amylase producers and the rhizobacteria are not, the production of this enzyme by the actinobacteria may have influenced the production of other metabolites that prevented the rhizobacteria from growing. Exceptions can be explained by the production of secondary metabolites by the actinobacterium itself (BOUBEKRI *et al.*, 2022) or by the biochemical apparatus of the rhizobium (FANG *et al.*, 2020).

Low phylogenetic proximity between two groups in coculture, such as the actinobacteria and rhizobacteria used in this work, means that these microorganisms are less likely to have overlapping growth requirements. Consequently, the low kinship between donor and recipient reduces, but does not eliminate, the extent of competition for resources between the co-inoculated bacteria. This facilitates the growth of the intended group (MITRI; FOSTER, 2013).

One way in which two bacteria cooperate is through the synthesis of *public goods*. These molecules are metabolites whose synthesis is very expensive for the cell, but which are released (actively or passively) into the environment, from where other microorganisms ("cheaters") can take advantage. These metabolites range from siderophores, biosurfactants, toxins, or exoenzymes capable of degrading polymers into smaller fragments absorbable by other bacteria (SMITH; SCHUSTER, 2019; BRUCE; WEST; GRIFFIN, 2019). Using the classification of Smith *et al.* (2019), this last case perfectly describes an example of substrate *cross-feeding*, where one strain feeds on the extracellular metabolic products of another. Substrate *cross-feeding* seems to be exactly what happened in our work. The exoenzymes, mainly cellulase and xylanase, which are *public goods* produced by the actinobacteria, were able to hydrolyze the respective substrates into simpler sugars that served as a carbon source for the growth of the rhizobacteria. However, according to D'Souza *et al.* (2018), the sharing of *public goods* occurs mainly between genealogically close individuals, due to the possibility of the producer of these metabolites spreading its genes indirectly. Given the already highlighted morphological, physiological and taxonomic dissimilarities between actinobacteria and rhizobacteria, the sharing of *public goods* between these microorganisms is an intriguing result, to say the least.

5.3 Selecting the most promising strains

Based on the CIs, the actinobacteria were distributed into 7 groups in terms of their similarity. Of these groups, the one containing actinobacteria A139 and A145 had a higher CI than the others. This result is illustrated in Figure 7, where the dendrogram resulting from the statistical analyses is shown alongside a *heatmap* plotted from their compatibility indices. The *heatmap* allows better observation of the CIs of each strain in each culture medium, illustrated by the shade of the colors.

FIGURE 7 - *Heatmap* of the actinobacteria compatibility index (ICA) in relation to rhizobacteria from the northeastern semi-arid region associated with cluster analysis

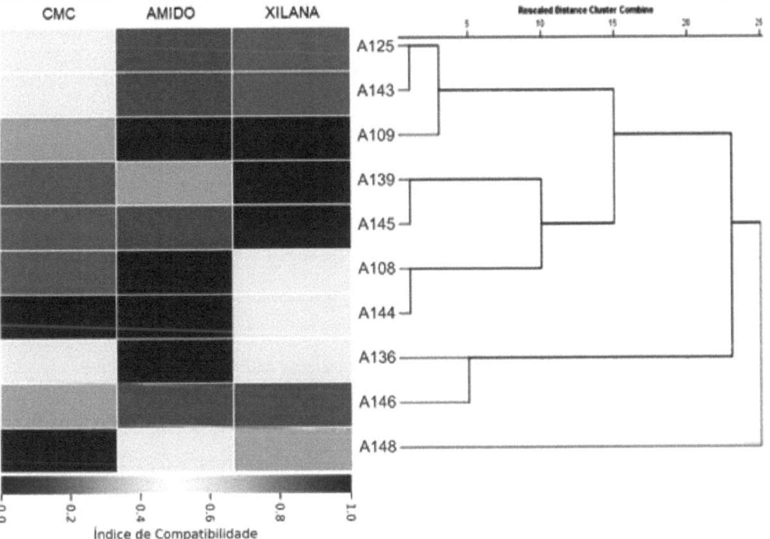

On the left of the image, you can see the *heatmap* with the actinobacteria compatibility indices. The x-axis shows the strains of actinobacteria used and the y-axis shows the culture media. On the right is the dendrogram obtained from the cluster analysis of the compatibility and antagonism indices. The actinobacteria were divided into 7 groups based on their similarity.
Source: the author

Bioinoculants in agriculture have various functions, including inhibiting the growth of phytopathogens, producing siderophores, fixing nitrogen, producing phytohormones and solubilizing phosphate. In addition, these inoculants are also capable of reducing the damage caused by chemical fertilizers (CHAUDHARY; SHUKLA, 2020). Several microorganisms are already used as biofertilizers, such as the genera *Pseudomonas, Bacillus, Phyllobacterium* and *Rhodococcus, as* well as rhizobia such as *Azorhizobium, Sinorhizobium* and *Bradyrhizobium,* and actinobacteria such as *Mycobacterium, Frankia, Arthrobacter* and *Streptomyces* (MAHANTY *et al.*, 2016).

FIGURE 8 - *Heatmap* of rhizobial compatibility indices (RCI) in relation to actinobacteria from the northeastern semi-arid region.

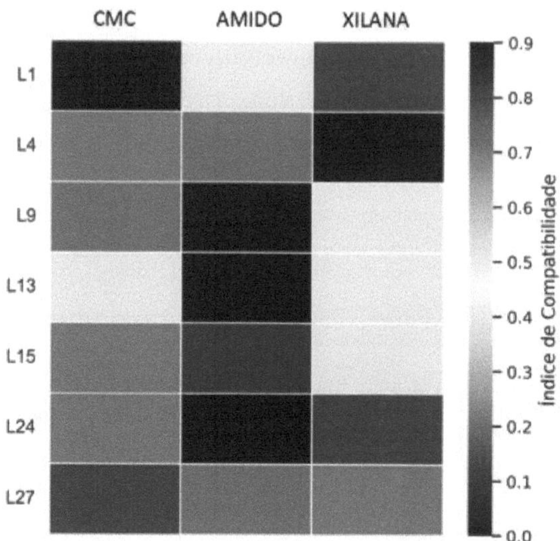

The x-axis shows the strains of rhizobia used and the y-axis shows the
culture media.
Source: the author

Research conducted by Htwe *et al.* (2019) showed that a biofertilizer based
on *Bradyrhizobium japonicum* SAY3-7, *Bradyrhizobium elkanii* BLY3-8 and
Streptomyces griseoflavus P4 had positive effects on the growth of mung beans, cowpeas
and soybeans. Especially when inoculated together with mung beans and soybeans, this
bioinoculant improved growth, nodulation, nitrogen fixation, NPK absorption and seed
yield in an experiment carried out in a greenhouse. Considering the metabolic
compatibility presented *in vitro* between strains A139 and A145 (*Streptomyces* sp.), and
L1 (*Bradyrhizobium elkanii*), it is clear that *in vivo* co-inoculation is necessary *in* order
to advance in the prospecting of this bioinoculant. It is possible to say, however, that *in
vitro co-inoculation* is an important step to eliminate possible antagonistic pairs before
proceeding to *in vivo* experiments. In addition, the research conducted here on growth-
promoting rhizobacteria isolated from the semi-arid region has provided an overview of
the interactions between these microorganisms from an ecological point of view.

Based on the antagonism index (AI) between the strains, two *heatmaps* were
also plotted to illustrate this data. For the CI, the colors ranged from red (lowest CI) to
blue (highest CI), since we wanted a CI closer to 1. However, for the AIs this is reversed.

42

As an AI closer to 0 is desired, the colors in these *heatmaps ranged* from blue (lowest AI) to red (highest AI).

FIGURE 9 - *Heatmaps of* the antagonism indices of actinobacteria (A) and rhizobia (B)

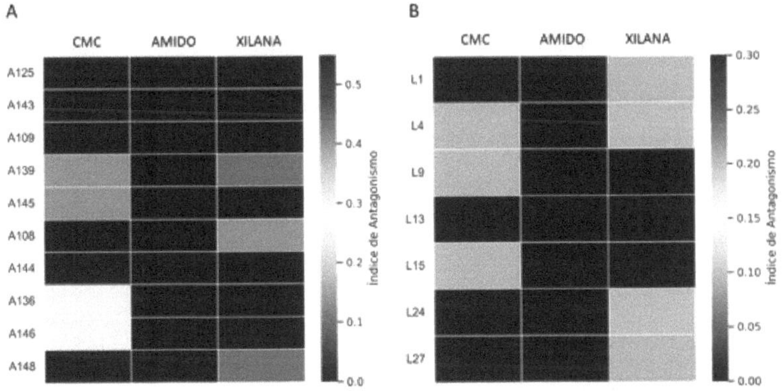

The x-axis shows the strains of actinobacteria (A) and rhizobacteria (B) used, and the y-axis shows the culture media.
Source: the author.

Despite the limited understanding of the mechanisms of action of various bacteria, studies indicate that the application of different microorganisms together allows for better growth and effective plant protection. The prospecting of these microorganisms must be multicomponent due to the considerable variability of the environment in order to produce optimum degrees of cooperation between the microorganism and the desired final effect on the plant. The combination of bacterial strains in a single formula improves efficacy and reliability, allowing for greater culture specificity and is highly promising (ZARDAK *et al.*, 2018; KOUR *et al.*, 2022).

Although the production of exoenzymes is necessary to colonize legume roots and carry out SNF, it is not common for rhizobia to have this metabolic apparatus, highlighting the importance of a microorganism that facilitates their growth in the presence of carbon sources that the rhizobium cannot assimilate. Co-inoculation with an actinobacterium from the *Streptomyces* genus is highly promising for the successful growth of rhizobia in these circumstances and consequently greater nitrogen fixation efficiency, reducing the demand for nitrogen fertilizers.

5.4 Scanning electron microscopy

Among Gram-positive bacteria, actinobacteria show the greatest morphological differentiation. These filamentous microorganisms form hyphae between 0.4 and 1.2 µm and spores between 1 and 2 µm thick, making these structures easily observable via SEM. The strains that were subjected to SEM (A139 and A145) showed similar morphological characteristics, with spiral spore chains (*Spira*) and spores with an irregular rough surface. The images obtained by SEM (Figure 10) associated with the micromorphological characterization are sufficient evidence to infer with some certainty that these strains belong to the genus *Streptomyces* (LI *et al*., 2016).

FIGURE 10 - Scanning electron microscopy (SEM) of strains A139 and A145.

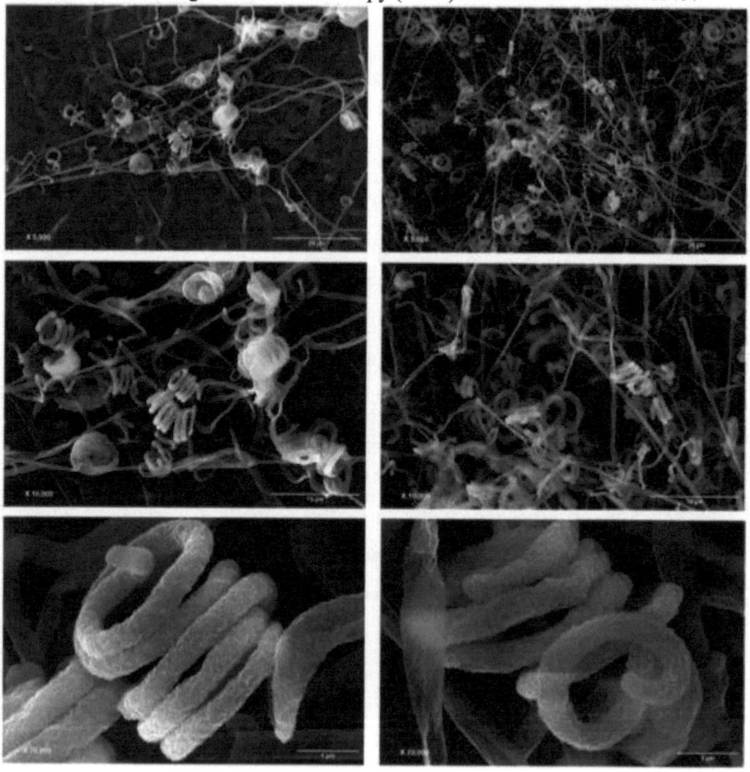

On the left are images of the actinobacterium A139, and on the right, A145. The first line shows the images at 5,000 X magnification, the second at 10,000 X and the third at 70,000 X. The characteristic spirals of actinobacteria of the *Streptomyces* genus are highlighted.
Source: the author

Scanning electron microscopy works, as the name suggests, by scanning high-energy electrons over the surface of a given sample. By having a much shorter wavelength than light, the electrons used in SEM are able to "differentiate" many more details and characteristics than an optical microscope, and can present resolutions better than 1nm and magnifications of 200,000 to 1,000,000X for more modern and powerful models (UL-HAMID, 2018). SEM is carried out in a vacuum chamber and can capture various signals that reveal information such as the morphology and composition of the sample being analyzed (AID *et al.*, 2020).

Actinobacteria isolated from arid and semi-arid regions have already been well studied and have shown great biotechnological potential, such as the control of phytopathogens, the production of biosurfactants and the production of enzymes (CASTAÑEDA-CISNEROS *et al.*, 2020). According to Parihar *et al.* (2022), the exploration of understudied extreme environments is interesting because of the possibility of obtaining new biomolecules and new species of actinobacteria. However, when it comes to genera, nature is more restricted. In their study on the diversity of actinobacteria isolated from saline basins in the Thar Desert in India, these authors found an abundance of 64% *Streptomyces*. This is not surprising, since this genus is actually the dominant one not only in abundance, but also in biotechnological importance (OLANREWAJU; BABALOLA, 2018). In the northeastern semi-arid region specifically, the situation is the same: *Streptomyces* is the most abundant genus of actinobacteria (BRITO *et al.*, 2015; SANTOS *et al.*, 2019b).

6 CONCLUSION

The actinobacteria were able to stimulate the growth of rhizobia in culture media not specific for diazotrophic bacteria. Strains A139 and A145 (*Streptomyces sp.*) showed the greatest compatibility with the rhizobacteria in the media tested. *Rhizobium* strain L1 (*Bradyrhizobium elkanii*) showed the best growth when co-inoculated with the actinobacteria using carboxymethylcellulose, starch and xylan as carbon sources, even without producing the enzymatic apparatus needed to degrade these carbohydrates.

Cooperative relationships between bacteria are common in nature, and the use of these interactions to improve the quality of life and food security of the population through biotechnology is promising. There are already bioinoculants being prospected from co-inoculations of *Streptomyces* and *Bradyrhizobium*, but due to the diversity and complexity of the bacteria, the plants, and the biotic and abiotic factors to which they are subjected, multifactorial bioprospecting of genetic potential at a local level is necessary. The results obtained in this study are an important step towards the development of a biofertilizer adapted to semi-arid soils, while at the same time valuing local microbial diversity.

7 FUTURE PROSPECTS

In vivo studies using combinations of the A139 and A145 actinobacteria strains with the L1 rhizobium strain would be important to validate the results when associated with plants. As it is a more complex organism, there is a chance that the plant will mediate the interactions between the microorganisms and that new and unexpected results will be obtained. It is clear, however, that a co-inoculation of *Streptomyces* and *Bradyrhizobium* has great potential for the development of a biofertilizer.

REFERENCES

ABE, M. M. *et al.* Production and assessment of the biodegradation and ecotoxicity of xylan- and starch-based bioplastics. **Chemosphere**, [*s. l.*.], v. 287, p. 132290, jan. 2022. http://dx.doi.org/10.1016/j.chemosphere.2021.132290

AID, S. R. *et al.* A Study on biological samples preparation for high resolution imaging of scanning electron microscope. **Journal of Physics**: Conference Series, [*s. l.*], v. 1447, n. 1, p. 012034, jan. 2020. http://dx.doi.org/10.1088/1742-6596/1447/1/012034

AL-AGAMY, M. H. *et al.* Production and partial characterization of α-amylase enzyme from marine actinomycetes. **Biomed Research International**, [*s. l.*], v. 2021, p. 1-15, dec. 2021. http://dx.doi.org/10.1155/2021/5289848

ALKAHTANI, M. D. F. *et al.* Isolation and characterization of plant growth promoting endophytic bacteria from desert plants and their application as bioinoculants for sustainable agriculture. **Agronomy**, [*s. l.*], v. 10, n. 9, p. 1325, sep. 2020. https://doi.org/10.3390/agronomy10091325

AL-SHAIBANI, M. M. *et al.* Biodiversity of secondary metabolites compounds isolated from phylum Actinobacteria and its therapeutic applications. **Molecules**, [*s. l.*], v. 26, n. 15, p. 4504, jul. 2021. http://dx.doi.org/10.3390/agronomy10091325

ANANDAN, R. *et al.* An Introduction to Actinobacteria. *In:* DHARUMADURAI, Dhanasekaran; **YI, Jiang. (Edit.) Actinobacteria - Basics and biotechnological applications.** [*S. l.*], p. 3-38, 2016. http://dx.doi.org/10.5772/62329

ARAUJO, R. *et al.* Biogeography and emerging significance of actinobacteria in Australia and Northern Antarctica soils. **Soil Biology and Biochemistry**, [*s. l.*], v. 146, p. 107805, jul. 2020. http://dx.doi.org/10.1016/j.soilbio.2020.107805

ATIENO, M. *et al.* Assessment of biofertilizer use for sustainable agriculture in the Great Mekong Region. **Journal of Environmental Management**, [*s. l.*], v. 275, p. 111300, dec. 2020. http://dx.doi.org/10.1016/j.jenvman.2020.111300

AXENOV-GIBANOV, D. V. *et al.* A. Actinobacteria isolated from an underground lake and moonmilk speleothem from the biggest conglomeratic karstic cave in Siberia as sources of novel biologically active compounds. **Plos One**, [*s. l.*], v. 11, n. 2, p. 1-21, feb. 2016. http://dx.doi.org/10.1371/journal.pone.0149216

BALAGURUNATHAN, R. *et al.* Characterization and Identification of Actinobacteria. In: BALAGURUNATHAN, R. *et al.* **Protocols in Actinobacterial Research**. New York: Springer, 2020. Chap. 3. p. 39-64. (Springer Protocols Handbooks). http://dx.doi.org/10.1007/978-1-0716-0728-2_3

BAO, Y. *et al.* Important ecophysiological roles of non-dominant actinobacteria in plant residue decomposition, especially in less fertile soils. **Microbiome**, [*s. l.*], v. 9, n. 1, p. 1-17, Apr. 2021. http://dx.doi.org/10.1186/s40168-021-01032-x

BARKA, E. A. *et al*. Taxonomy, Physiology, and natural products of actinobacteria. **Microbiology And Molecular Biology Reviews**, [*s. l.*], v. 80, n. 1, p. 1-43, nov. 2015. http://dx.doi.org/10.1128/mmbr.00019-15

BASILE, L. A.; LEPEK, Viviana C. Legume-rhizobium dance: an agricultural tool that could be improved? **Microbial Biotechnology**, [*s. l.*], v. 14, n. 5, p. 1897-1917, jul. 2021. http://dx.doi.org/10.1111/1751-7915.13906

BASU, A. *et al*. Plant growth promoting rhizobacteria (PGPR) as green bioinoculants: recent developments, constraints, and prospects. **Sustainability**, [*s. l.*], v. 13, n. 3, p. 1140, jan. 2021. http://dx.doi.org/10.3390/su13031140

BEHIE, S. W. *et al*. Molecules to ecosystems: actinomycete natural products *in situ*. **Frontiers In Microbiology**, [*s. l.*], v. 7, p. 1-11, jan. 2017. http://dx.doi.org/10.3389/fmicb.2016.02149

BHATI, N. *et al*. Cost-effective cellulase production, improvement strategies, and future challenges. **Journal of Food Process Engineering**, [*s. l.*], v. 44, n. 2, p. 1-11, dec. 2020. http://dx.doi.org/10.1111/jfpe.13623.

BOSSE, M. A. *et al*. Physiological impact of flavonoids on nodulation and ureide metabolism in legume plants. **Plant Physiology and Biochemistry**, [*s. l.*], v. 166, p. 512-521, set. 2021. http://dx.doi.org/10.1016/j.plaphy.2021.06.007

BOUBEKRI, K. *et al*. Multifunctional role of Actinobacteria in agricultural production sustainability: a review. **Microbiological Research**, [S.L.], v. 261, p. 127059, Aug. 2022. http://dx.doi.org/10.1016/j.micres.2022.127059

BOUKAR, O. *et al*. Cowpea (*Vigna unguiculata*): genetics, genomics and breeding. **Plant Breeding**, [*s. l.*], v. 138, n. 4, p. 415-424, May 2018. http://dx.doi.org/10.1111/pbr.12589

BRITO, F. A. E. *et al*. Rhizospheric soil actinobacteria in the caatinga biome. **Enciclopédia Biosfera**, Goiânia - Go, v. 11, n. 21, p. 1992-2004, jun. 2015.

BROWN-ELLIOTT, B. A. *et al*. Current Status of *Nocardia* taxonomy and recommended identification methods. **Clinical Microbiology Newsletter**, [*s. l*], v. 37, n. 4, p. 25-32, feb. 2015. https://doi.org/10.1016/j.clinmicnews.2015.01.007

BRUCE, J. B. *et al*. Functional amyloids promote retention of public goods in bacteria. **Proceedings Of The Royal Society B**: Biological Sciences, [*s. l.*], v. 286, n. 1903, p. 20190709, 29 May 2019. http://dx.doi.org/10.1098/rspb.2019.0709

CAI, P. *et al*. Soil biofilms: microbial interactions, challenges, and advanced techniques for *ex-situ* characterization. **Soil Ecology Letters**, [*s. l.*], v. 1, n 3-4, p. 85-93, jul. 2019. http://dx.doi.org/10.1007/s42832-019-0017-7

CASTAÑEDA-CISNEROS, Y. E. *et al.* Isolation and selection of *Streptomyces* species from semi-arid agricultural soils and their potential as producers of xylanases and cellulases. **Current Microbiology**, [*s. l.*], v. 77, n. 11, p. 3460-3472, aug. 2020. http://dx.doi.org/10.1007/s00284-020-02160-7

CHAUDHARY, T. *et al.* Techniques for improving formulations of bioinoculants. **3 Biotech**, [*s. l.*], v. 10, n. 5, p. 1-9, abr. 2020. http://dx.doi.org/10.1007/s13205-020-02182-9

CHAUDHARY, T.; SHUKLA, P. Bioinoculants for bioremediation applications and disease resistance: innovative perspectives. **Indian Journal of Microbiology**, [*s. l.*], v. 59, n. 2, p. 129-136, feb. 2019. http://dx.doi.org/10.1007/s12088-019-00783-4

ÇİÇEKLER, M. Biopulping and biobleaching in pulp and paper industry: a review of xylanase and its effects. *In*: BERAM, A.; ULUSAN, M. D. **Forest and agricultural studies from different perspectives**. Lithuania: Sra Academic Publishing, 2022. Chap. 2. p. 21-42

CLÚA, J.; *et al.* Compatibility between legumes and rhizobia for the establishment of a successful nitrogen-fixing symbiosis. **Genes**, [*s. l.*], v. 9, n. 3, p. 125, feb. 2018. http://dx.doi.org/10.3390/genes9030125

DAS, R. *et al.* Antimicrobial potentiality of actinobacteria isolated from two microbiologically unexplored forest ecosystems of Northeast India. **Bmc Microbiology**, [s.l.], v. 18, n. 1, p. 71, 11 jul. 2018. http://dx.doi.org/10.1186/s12866-018-1215-7

DICENZO, G. C. *et al.* Multidisciplinary approaches for studying rhizobium-legume symbioses. **Canadian Journal of Microbiology**, [*s. l.*], v. 65, n. 1, p. 1-33, sep. 2018. http://dx.doi.org/10.1139/cjm-2018-0377

DINESHKUMAR, R. *et al.* Microalgae as bio-fertilizers for rice growth and seed yield productivity. **Waste and Biomass Valorization**, [*s. l.*], v. 9, n. 5, p. 793-800, feb. 2017. http://dx.doi.org/10.1007/s12649-017-9873-5

DORNELAS, J. C. M. *et al.* Characterization and phylogenetic affiliation of actinobacteria from tropical soils with potential uses for agro-industrial processes. **Genetics And Molecular Research**, [*s. l.*], v. 16, n. 3, p. 1, 2017. http://dx.doi.org/10.4238/gmr16039703

D'SOUZA, G. *et al.* Ecology and evolution of metabolic cross-feeding interactions in bacteria. **Natural Product Reports**, [*s. l.*], v. 35, n. 5, p. 455-488, 2018. http://dx.doi.org/10.1039/c8np00009c

DUTTA, S.; PODILE, A. R. Plant Growth Promoting Rhizobacteria (PGPR): the bugs to debug the root zone. **Critical Reviews In Microbiology**, [*s. l.*], v. 36, n. 3, p. 232-244, jul. 2010. http://dx.doi.org/10.3109/10408411003766806

ENEBE, M. C.; BABALOLA, O. O. The influence of plant growth-promoting rhizobacteria in plant tolerance to abiotic stress: a survival strategy. **Applied Microbiology and Biotechnology**, [*s. l.*], v. 102, n. 18, p. 7821-7835, jul. 2018. http://dx.doi.org/10.1007/s00253-018-9214-z

EVANS, C. R. *et al*. Metabolic heterogeneity and cross-feeding in bacterial multicellular systems. **Trends In Microbiology**, [*s. l.*], v. 28, n. 9, p. 732-743, sep. 2020. http://dx.doi.org/10.1016/j.tim.2020.03.008

FANG, L. *et al*. Exogenous application of signaling molecules to enhance the resistance of legume-rhizobium symbiosis in Pb/Cd-contaminated soils. **Environmental Pollution**, [s.l.], v. 265, p. 114744, Oct. 2020. http://dx.doi.org/10.1016/j.envpol.2020.114744

FATMAWATI, U. *et al*. Isolation of actinomycetes from maize rhizosphere from Kupang, east Nusa Tenggara province, and evaluation of their antibacterial, antifungal, and extracellular enzyme activity. **Indonesian Journal Of Biotechnology**, [*s. l.*], v. 23, n. 1, p. 40, jun. 2018. http://dx.doi.org/10.22146/ijbiotech.33064

FRANCO-CORREA, M.; CHAVARRO-ANZOLA, V. Actinobacteria as plant growth-promoting rhizobacteria. **Actinobacteria - basics and biotechnological Applications**, [*s. l.*], p. 249-270, feb. 2016. http://dx.doi.org/10.5772/61291

GEORGE, M. *et al*. Distribution and bioactive potential of soil actinomycetes from different ecological habitats. **African Journal of Microbiology Research**, [*s. l.*], v. 6, n. 10, p. 2265-2271, mar. 2012. http://dx.doi.org/10.5897/ajmr11.856

GIRI, Samir *et al*. Metabolic dissimilarity determines the establishment of cross-feeding interactions in bacteria. **Current Biology**, [s.l.], v. 31, n. 24, p. 5547-5557, dec. 2021. http://dx.doi.org/10.1016/j.cub.2021.10.019

GIRIJA, D. *et al*. Isolation and characterization of native cowpea rhizobia from Wayanad India. **Legume Research**, [*s. l*], v. 43, n. 1, p. 126-133, nov. 2018.

GOBALAKRISHNAN, R. Phylogenctic diversity of culturable marine actinobacteria isolated from the Havelock island, the Andamans, India. **Ecological Genetics And Genomics**, [*s. l.*], v. 23, p. 100123, jun. 2022. http://dx.doi.org/10.1016/j.egg.2022.100123

GOODFELLOW, M. *et al*. Rare taxa and dark microbial matter: novel bioactive actinobacteria abound in Atacama desert soils. **Antonie van Leeuwenhoek**, [*s. l.*], v. 111, n. 8, p. 1315-1332, May 2018. http://dx.doi.org/10.1007/s10482-018-1088-7

GRAM, C. Ueber die isolirte Farbung der Schizomyceten in Schnitt-und Trockenpraparaten. **Fortschritte der Medicin**, v. 2, p. 185-189, 1884.

HAMEDI, J.; POORINMOHAMMAD, N. The Cellular structure of actinobacteria *In:* WINK, J. *et al*. **Biology and biotechnology of actinobacteria**, [*s. l.*]: Springer International Publishing. 2017, p. 5-28. http://dx.doi.org/10.1007/978-3-319-60339-1_2

HAMID, M. E. *et al.* Diversity and geographic distribution of soil streptomycetes with antagonistic potential against actinomycetoma-causing *Streptomyces sudanensis* in Sudan and South Sudan. **Bmc Microbiology**, [*s. l.*], v. 20, n. 1, p. 1-13, feb. 2020. http://dx.doi.org/10.1186/s12866-020-1717-y

HAYAT, S.; *et al.* Actinobacteria: potential candidates as plant growth promoters. **Plant Stress Physiology**, [*s. l.*], p. 1-17, Sep. 2020. http://dx.doi.org/10.5772/intechopen.93272

HAZARIKA, S. N.; THAKUR, D. Actinobacteria. *In*: AMARESAN, N. *et al.* **Beneficial Microbes in Agro-Ecology**, [*s. l.*]: Elsevier. 2020, p. 443-476. Chapter 21. http://dx.doi.org/10.1016/b978-0-12-823414-3.00021-6

HELENE, L. C. F. *et al.* New insights into the taxonomy of bacteria in the genomic era and a case study with rhizobia. **International Journal of Microbiology**, [*s. l.*], v. 2022, p. 1-19, May, 2022. Hindawi Limited. http://dx.doi.org/10.1155/2022/4623713

HTWE, A. Z. *et al.* Effects of co-inoculation of *Bradyrhizobium japonicum* SAY3-7 and *Streptomyces griseoflavus* P4 on plant growth, nodulation, nitrogen fixation, nutrient uptake, and yield of soybean in a field condition. **Soil Science And Plant Nutrition**, [*s. l.*], v. 64, n. 2, p. 222-229, jan. 2018. http://dx.doi.org/10.1080/00380768.2017.1421436

HTWE, A. Z. *et al.* Effects of biofertilizer produced from *Bradyrhizobium* and *Streptomyces griseoflavus* on plant growth, nodulation, nitrogen fixation, nutrient uptake, and seed yield of mung bean, cowpea, and soybean. **Agronomy**, [*s. l.*], v. 9, n. 2, p. 77, feb. 2019. http://dx.doi.org/10.3390/agronomy9020077

HTWE, A. Z.; YAMAKAWA, T. Low-Density Co-Inoculation with *Bradyrhizobium japonicum* SAY3-7 and *Streptomyces griseoflavus* P4 promotes plant growth and nitrogen fixation in soybean cultivars. **American Journal of Plant Sciences**, [*s. l.*], v. 7, n. 12, p. 1652-1661, 2016. http://dx.doi.org/10.4236/ajps.2016.712156

ISLAM, F.; ROY, N. Isolation and characterization of cellulase-producing bacteria from sugar industry waste. **American Journal of Bioscience,** [*s. l.*], v. 7, n. 1, p. 16, 2019. http://dx.doi.org/10.11648/j.ajbio.20190701.13

JABBOROVA, D.; *et al.* Co-inoculation of rhizobacteria promotes growth, yield, and nutrient contents in soybean and improves soil enzymes and nutrients under drought conditions. **Scientific Reports**, [*s. l.*], v. 11, n. 1, p. 1, nov. 2021. http://dx.doi.org/10.1038/s41598-021-01337-9

JI, J. *et al.* Genome editing in Cowpea *Vigna unguiculata* Using CRISPR-Cas9. **International Journal of Molecular Sciences**, [*s. l.*], v. 20, n. 10, p. 2471, May 2019. http://dx.doi.org/10.3390/ijms20102471

JOSE, P. A. *et al.* Actinobacteria in natural products research: progress and prospects. **Microbiological Research,** [*s. l.*], v. 246, p. 126708, jan 2021. http://dx.doi.org/10.1016/j.micres.2021.126708

JOSE, P. A. *et al*. *Non-Streptomyces* actinomycetes and natural products: recent updates. **Studies in Natural Products Chemistry**, [*s. l.*], p. 395-409, 2019. http://dx.doi.org/10.1016/b978-0-444-64183-0.00011-7

KAUSHAL, J. *et al*. A multifaceted enzyme conspicuous in fruit juice clarification: an elaborate review on xylanase. **International Journal of Biological Macromolecules**, [*s. l.*], v. 193, p. 1350-1361, dec. 2021. http://dx.doi.org/10.1016/j.ijbiomac.2021.10.194

KLEIN, T. A. *et al*. Contact-Dependent interbacterial antagonism mediated by protein secretion machines. **Trends In Microbiology**, [*s. l.*], v. 28, n. 5, p. 387-400, May. 2020. http://dx.doi.org/10.1016/j.tim.2020.01.003

KOUR, D. *et al*. Drought adaptive microbes as bioinoculants for the horticultural crops. **Heliyon**, [s.l.], v. 8, n. 5, p. 9493, May 2022. http://dx.doi.org/10.1016/j.heliyon.2022.e09493

KUMAR, M. *et al*. Potential applications of extracellular enzymes from *Streptomyces* spp. in various industries. **Archives of Microbiology**, [*s. l.*], v. 202, n. 7, p. 1597-1615, May 2020. http://dx.doi.org/10.1007/s00203-020-01898-9

KUMAR, V. V. Biofertilizers and biopesticides in sustainable agriculture. **Role of Rhizospheric Microbes In Soil**, [*s. l.*], p. 377-398, 2018. http://dx.doi.org/10.1007/978-981-10-8402-7_14

KUMAWAT, K. C. *et al*. Co-inoculation of indigenous *Pseudomonas oryzihabitans* and *Bradyrhizobium* sp. modulates the growth, symbiotic efficacy, nutrient acquisition, and grain yield of soybean. **Pedosphere**, [*s. l.*], v. 32, n. 3, p. 438-451, jun. 2022. http://dx.doi.org/10.1016/s1002-0160(21)60085-1

KURTBÖKE, D. I. *et al*. Actinobacteria in Marine Environments. *In:* KIM, S. K. **Encyclopedia of Marine Biotechnology**: five volume set. [S.L.]: John Wiley & Sons. 2020. Section VIII, Microbiology, chap. 86. p. 1951-1978. https://doi.org/10.1002/9781119143802.ch86

KUYKENDALL, L. D. *et al*. *Rhizobium*. **Bergey'S Manual of Systematics of Archaea and bacteria**, [s.l.]: Wiley. 2015. p. 1-6 http://dx.doi.org/10.1002/9781118960608.gbm00847

KWON, H. Y. *et al*. Development of a universal fluorescent probe for Gram-Positive Bacteria. **Angewandte Chemie**, [*s. l.*], p. 8514-8519, May. 2019. http://dx.doi.org/10.1002/ange.201902537

LACOMBE-HARVEY, M. È. *Et al*. Chitinolytic functions in actinobacteria: ecology, enzymes, and evolution. **Applied Microbiology and Biotechnology**, [*s. l.*], v. 102, n. 17, p. 7219-7230, jun. 2018. http://dx.doi.org/10.1007/s00253-018-9149-4

LADAN, W. H. *et al*. ROLE OF PLANT GROWTH PROMOTING RHIZOBIA

STRAINS IN AGRICULTURE FOR SUSTAINABLE CROP YIELD (A REVIEW). **Bayero Journal Of Pure And Applied Sciences**, [*s. l*], v. 13, n. 1, p. 326-335, Apr. 2022.

LARSBRINK, J.; MCKEE, L. S. Bacteroidetes bacteria in the soil: glycan acquisition, enzyme secretion, and gliding motility. **Advances In Applied Microbiology**, [*s. l.*], p. 63-98, 2020. http://dx.doi.org/10.1016/bs.aambs.2019.11.001

LAU, E. T. *et al.* Plant growth-promoting bacteria as potential bio-inoculants and biocontrol agents to promote black pepper plant cultivation. **Microbiological Research**, [*s. l.*], v. 240, p. 126549, nov. 2020. http://dx.doi.org/10.1016/j.micres.2020.126549

LAW, J. W. F. *et al.* A Review on mangrove actinobacterial diversity: the roles of *Streptomyces* and novel species discovery. **Progress in Microbes and Molecular Biology**, [*s. l*], v. 2, n. 1, p. 1-10, May. 2019. DOI:10.36877/pmmb.a0000024

LAW, J. W. F. *et al.* The Rising of modern Actinobacteria era. **Progress in Microbes and Molecular Biology**, [*s. l*], v. 3, n. 1, p. 1- 6, 2020. https://doi.org/10.36877/pmmb.a0000064

LEWIN, G. R. *et al.* Evolution and ecology of actinobacteria and their bioenergy applications. **Annual Review of Microbiology**, [*s. l.*], v. 70, n. 1, p. 235-254, sep. 2016. http://dx.doi.org/10.1146/annurev-micro-102215-095748

LI, Q. *et al.* Morphological identification of actinobacteria. **Actinobacteria - Basics and Biotechnological Applications**, [*s. l.*], p. 59-86, feb. 2016. http://dx.doi.org/10.5772/61461

LIMA, J. V. L. *et al.* Characterization of actinobacteria from the semiarid region, and their antagonistic effect on strains of rhizobia. **African Journal of Biotechnology**, [*s. l.*], v. 16, n. 11, p. 499-507, mar. 2017. http://dx.doi.org/10.5897/ajb2016.15724

LINDSTRÖM, K.; MOUSAVI, S. A. Effectiveness of nitrogen fixation in rhizobia. **Microbial Biotechnology**, [*s. l.*], v. 13, n. 5, p. 1314-1335, dec. 2019. http://dx.doi.org/10.1111/1751-7915.13517

LITTLE, A. E. F. *et al.* Rules of Engagement: interspecies interactions that regulate microbial communities. **Annual Review of Microbiology**, [*s. l.*], v. 62, n. 1, p. 375-401, Oct. 2008. http://dx.doi.org/10.1146/annurev.micro.030608.101423

MAHANTY, T. *et al.* Biofertilizers: a potential approach for sustainable agriculture development. **Environmental Science and Pollution Research**, [s.l.], v. 24, n. 4, p. 3315-3335, nov. 2016. http://dx.doi.org/10.1007/s11356-016-8104-0

MAITRA, S. *et al.* Bioinoculants-Natural biological resources for sustainable plant production. **Microorganisms**, [*s. l.*], v. 10, n. 1, p. 51, dec. 2021. http://dx.doi.org/10.3390/microorganisms10010051

MAJIDZADEH, M. *et al.* Antimicrobial activity of Actinobacteria isolated from dry land soil in Yazd, Iran. **Molecular Biology Reports**, [*s. l.*], v. 48, n. 2, p. 1717-1723, feb. 2021. http://dx.doi.org/10.1007/s11033-021-06218-y

MANDAL, A. *et al.* Impact of genetically modified crops on rhizosphere microorganisms and processes: a review focusing on Bt cotton. **Applied Soil Ecology**, [*s. l.*], v. 148, p. 103492, Apr. 2020. http://dx.doi.org/10.1016/j.apsoil.2019.103492

MANFREDINI, A. *et al.* Current methods, common practices, and perspectives in tracking and monitoring bioinoculants in soil. **Frontiers in Microbiology**, [*s. l.*], v. 12, p. 1-22, Aug. 2021. http://dx.doi.org/10.3389/fmicb.2021.698491

MEDEIROS, E. *et al.* Cultural diversity of actinobacterial strains from the semi-arid region. **Enciclopédia Biosfera**, [*s. l.*], v. 15, n. 27, p. 205-218, jun. 2018. http://dx.doi.org/10.18677/encibio_2018a87

MESQUITA, A. F. N. *et al.* Amylase-Producing Actinobacteria Facilitate Rhizobia Growth in a Culture Medium with Starch. **World Wide Journal Of Multidisciplinary Research And Development**, [S.L.], v. 8, n. 11, p. 91-94, nov. 2022. http://dx.doi.org/10.5281/zenodo.7406324

MIA, M. A. B.; SHAMSUDDIN, Z. H. *Rhizobium* as a crop enhancer and biofertilizer for increased cereal production. **African Journal of Biotechnology**, [*s. l*], p. 6001-6009, Sep. 13, 2010

MIHAJLOVSKI, K. *et al.* From Agricultural Waste to Biofuel: enzymatic potential of a bacterial isolate *Streptomyces fulvissimus* CKS7 for bioethanol production. **Waste And Biomass Valorization**, [S.L.], v. 12, n. 1, p. 165-174, 10 feb. 2020. http://dx.doi.org/10.1007/s12649-020-00960-3

MILJAKOVIĆ, D. *et al.* Bio-priming of soybean with *Bradyrhizobium japonicum* and *Bacillus megaterium*: strategy to improve seed germination and the initial seedling growth. **Plants**, [*s. l.*], v. 11, n. 15, p. 1927, jul. 2022. http://dx.doi.org/10.3390/plants11151927

MITRI, S.; FOSTER, K. R. The Genotypic View of Social Interactions in Microbial Communities. **Annual Review Of Genetics**, [S.L.], v. 47, n. 1, p. 247-273, Nov. 23, 2013. http://dx.doi.org/10.1146/annurev-genet-111212-133307

NAYAK, S. K. *et al.* Rhizobacteria and its biofilm for sustainable agriculture: a concise review. *In*: YADAV, M. K.; SINGH, B. P. **New and future developments in microbial biotechnology and bioengineering: Microbial Biofilms**, [*s.l.*]: Elsevier. 2020, p. 165-175. http://dx.doi.org/10.1016/b978-0-444-64279-0.00013-x

NOBBE, F.; HILTNER, L. **INOCULATION OF THE SOIL FOR CULTIVATING LEGUMINOUS PLANTS**. Owner: United States Patent Office. USA n. 570813. Filing: Aug. 9, 1895. Grant: November 3, 1896.

NOFIANI, R. *et al.* Antibacterial Activities and Toxicity of *Streptosporangium* sp. SM1P. **Molekul**, [S.L.], v. 16, n. 3, p. 210, Nov. 15, 2021. http://dx.doi.org/10.20884/1.jm.2021.16.3.780

OLANREWAJU, O. S.; BABALOLA, O. O.. *Streptomyces*: implications and interactions in plant growth promotion. **Applied microbiology and biotechnology**, [*s. l.*], v. 103, n. 3, p. 1179-1188, dec. 2018. http://dx.doi.org/10.1007/s00253-018-09577-y

OO, K. T. *et al.* Isolation, Screening and Molecular Characterization of Multifunctional Plant Growth Promoting Rhizobacteria for a Sustainable Agriculture. **American Journal Of Plant Sciences**, [S.L.], v. 11, n. 06, p. 773-792, jun. 2020. http://dx.doi.org/10.4236/ajps.2020.116055

OROZCO-MOSQUEDA, M. C. *et al.* Plant growth-promoting bacteria as bioinoculants: attributes and challenges for sustainable crop improvement. **Agronomy**, [*s. l.*], v. 11, n. 6, p. 1167, jun. 2021. http://dx.doi.org/10.3390/agronomy11061167

OSIPITAN, O. A. *et al.* Production systems and prospects of cowpea (*Vigna unguiculata* (L.) Walp.) in the United States. **Agronomy**, [*s. l.*], v. 11, n. 11, p. 2312, nov. 2021. http://dx.doi.org/10.3390/agronomy11112312

PACIOS-MICHELENA, S. *et al.* Application of *Streptomyces* Antimicrobial Compounds for the Control of Phytopathogens. **Frontiers In Sustainable Food Systems**, [S.L.], v. 5, p. 696518, 9 sep. 2021. http://dx.doi.org/10.3389/fsufs.2021.696518

PALAI, J. B. Role of *Rhizobium* on growth and development of groundnut: a review. **International Journal of Agriculture Environment and Biotechnology**, [*s.l.*], v. 14, n. 1, p. 63-75, mar. 2021. http://dx.doi.org/10.30954/0974-1712.01.2021.7

PANDE, S. *et al.* Metabolic cross-feeding via intercellular nanotubes among bacteria. **Nature Communications**, [*s. l.*], v. 6, n. 1, p. 1-13, feb. 2015. http://dx.doi.org/10.1038/ncomms7238

PARIHAR, K. *et al.* Species composition and diversity dynamics of actinomycetes in arid and semi-arid salt basins of Rajasthan. **Current Microbiology**, [*s.l.*], v. 79, n. 6, p. 168, Apr. 2022. http://dx.doi.org/10.1007/s00284-022-02851-3

PARTE, A. C. *et al.* List of Prokaryotic names with Standing in Nomenclature (LPSN) moves to the DSMZ. **International Journal of Systematic and Evolutionary Microbiology**, [*s. l.*], v. 70, n. 11, p. 5607-5612, nov. 2020. http://dx.doi.org/10.1099/ijsem.0.004332

PATHAK, D. V.; KUMAR, M. Microbial Inoculants as Biofertilizers and Biopesticides. **Microbial inoculants in sustainable agricultural productivity**, *[s.n.t.]*, p. 197-209, 2016. http://dx.doi.org/10.1007/978-81-322-2647-5_11

PHELAN, V. V. *et al.* Microbial metabolic exchange-the chemotype-to-phenotype link. **Nature Chemical Biology**, [*s. l.*], v. 8, n. 1, p. 26-35, dec. 2011. http://dx.doi.org/10.1038/nchembio.739

POLTURAK, G. *et al.* New and emerging concepts in the evolution and function of plant biosynthetic gene clusters. **Current Opinion in Green And Sustainable Chemistry**, [*s. l.*], v. 33, p. 100568, feb. 2022. http://dx.doi.org/10.1016/j.cogsc.2021.100568

PONOMAROVA, O.; PATIL, K. R. Metabolic interactions in microbial communities: untangling the gordian knot. **Current Opinion In Microbiology**, [S.L.], v. 27, p. 37-44, oct. 2015. http://dx.doi.org/10.1016/j.mib.2015.06.014

POOLE, P. *et al.* Rhizobia: from saprophytes to endosymbionts. **Nature Reviews Microbiology**, [*s. l.*], v. 16, n. 5, p. 291-303, jan. 2018. http://dx.doi.org/10.1038/nrmicro.2017.171

PRASAD, M. *et al.* Plant growth promoting Rhizobacteria (PGPR) for sustainable agriculture. **Pgpr Amelioration in Sustainable Agriculture,** [*s. l.*], p. 129-157, 2019. http://dx.doi.org/10.1016/b978-0-12-815879-1.00007-0

PUTRI, A L; SETIAWAN, R. Isolation and screening of actinomycetes producing cellulase and xylanase from Mamasa soil, West Sulawesi. **Iop Conference Series**: Earth and Environmental Science, [*s. l.*], v. 308, n. 1, p. 012035, Aug. 2019. http://dx.doi.org/10.1088/1755-1315/308/1/012035

RACHMANIA, M. K. *et al.* Xylan-degrading ability of thermophilic actinobacteria from soil in a geothermal area. **Biodiversitas Journal of Biological Diversity**, [*s. l.*], v. 21, n. 1, p. 144-154, jan. 2020. http://dx.doi.org/10.13057/biodiv/d210119

RAHLWES, K. C.; *et al.* Cell walls and membranes of actinobacteria. **Subcellular Biochemistry,** [*s. l.*], p. 417-469, 2019. http://dx.doi.org/10.1007/978-3-030-18768-2_13

RAJKUMARI, J. *et al.* The changing paradigm of rhizobial taxonomy and its systematic growth upto postgenomic technologies. **World Journal of Microbiology and Biotechnology**, [*s. l.*], v. 38, n. 11, p. 206, aug. 2022. http://dx.doi.org/10.1007/s11274-022-03370-w

RAMOS, K. A. *et al.* Characterization and chromogenic diversity of actinobacteria from a microbial niche preserved in the Caatinga biome. **Enciclopédia Biosfera**, Goiânia - GO, v. 11, n. 21, p. 2115-2125, jun. 2015.

RODRIGUES, T. F. *et al.* Impact of pesticides in properties of *Bradyrhizobium* spp. and in the symbiotic performance with soybean. **World Journal of Microbiology and Biotechnology**, [*s. l.*], v. 36, n. 11, p. 172, oct. 2020. http://dx.doi.org/10.1007/s11274-020-02949-5

ROHDE, M. The Gram-Positive Bacterial Cell Wall. **Microbiology Spectrum**, [*s. l.*], v. 7, n. 3, p. 1-21, May 2019. http://dx.doi.org/10.1128/microbiolspec.gpp3-0044-2018

SADOWSKY, M. J. *et al*. Biochemical Characterization of fast- and slow-growing rhizobia that nodulate soybeans. **International Journal of Systematic Bacteriology**, [*s. l*], v. 33, n. 4, p. 716-722, Oct. 1983.

SAIDI, S. *et al*. Improvement of *Medicago sativa* crops productivity by the co-inoculation of *Sinorhizobium meliloti-Actinobacteria* Under Salt Stress. **Current Microbiology**, [*s. l*], v. 78, n. 4, p. 1344-1357, mar. 2021. http://dx.doi.org/10.1007/s00284-021-02394-z

SALWAN, R.; SHARMA, V. Molecular and biotechnological aspects of secondary metabolites in actinobacteria. **Microbiological Research**, [*s. l*], v. 231, p. 126374, jan. 2020. http://dx.doi.org/10.1016/j.micres.2019.126374

SANTOS, F. *et al*. Cultural diversity of actinobacteria strains from areas susceptible to desertification. **Enciclopédia Biosfera**, [*s. l*], v. 16, n. 29, p. 1844-1856, jun. 2019a. http://dx.doi.org/10.18677/encibio_2019a142

SANTOS, F. *et al*. Morphology of actinobacteria strains in areas susceptible to desertification. **Enciclopédia Biosfera**, [*s. l*], v. 16, n. 29, p. 1911-1924, jun. 2019b. http://dx.doi.org/10.18677/encibio_2019a148

SATRIA, H. *et al*. Extracellular hydrolytic enzyme activities of indigenous actinomycetes on pretreated bagasse using choline acetate ionic liquid. **Biocatalysisa and Agricultural Biotechnology**, [*s. l*], v. 24, p. 101503, mar. 2020. http://dx.doi.org/10.1016/j.bcab.2020.101503

SCHNEIDER, S. *et al*. Sulfate is transported at significant rates through the symbiosome membrane and is crucial for nitrogenase biosynthesis. **Plant, Cell & Environment**, [*s. l*], v. 42, n. 4, p. 1180-1189, nov. 2018. http://dx.doi.org/10.1111/pce.13481

SCHULTE, C. C. M. *et al*. Metabolic control of nitrogen fixation in rhizobium-legume symbioses. **Science Advances**, [*s. l*], v. 7, n. 31, p. 2433, jul. 2021. http://dx.doi.org/10.1126/sciadv.abh2433

SHARMA, K. *et al*. Transparent soil microcosms for live-cell imaging and non-destructive stable isotope probing of soil microorganisms. **Elife**, [*s. l*], v. 9, p. 1-28, nov. 2020. http://dx.doi.org/10.7554/elife.56275

SHETEIWY, M. S. *et al*. Physiological and biochemical responses of soybean plants inoculated with Arbuscular mycorrhizal fungi and *Bradyrhizobium* under drought stress. **Bmc Plant Biology**, [*s. l*], v. 21, n. 1, p. 1-21, apr. 2021. http://dx.doi.org/10.1186/s12870-021-02949-z

SILVA, M. *et al*. DIVERSITY OF ACTINOBACTERIAL CELLS IN THE RPPN "FAZENDA NÃO ME DEIXES"-QUIXADÁ (CE). **Enciclopédia Biosfera**, [S.L.], v. 16, n. 29, p. 1857-1869, 30 jun. 2019b. http://dx.doi.org/10.18677/encibio_2019a143

SILVA, V. B. **Micro-organisms colonizing nodules of *Vigna* spp. grown in Caatinga soils**. Thesis. 2020. Federal University of Paraíba, Center for Agricultural Sciences - CCA. Graduate Program in Soil Science - PPGCS. Areia, PB, 2020

SILVA, V. M. A. Cross-feeding among soil bacterial populations: selection and characterization of potential bio-inoculants. **Journal of Agricultural Science**, [*s. l*], v. 11, n. 5, p. 23, abr. 2019a. http://dx.doi.org/10.5539/jas.v11n5p23

SINGH, R; DUBEY, A. K.. Diversity and applications of endophytic actinobacteria of plants in special and other ecological niches. **Frontiers in Microbiology**, [*s. l*], v. 9, p. 1767, Aug. 2018. http://dx.doi.org/10.3389/fmicb.2018.01767

SMITH, N. W. *et al.* The Classification and Evolution of Bacterial Cross-Feeding. **Frontiers in Ecology and Evolution**, [*s. l*], v. 7, p. 1-15, May. 2019. http://dx.doi.org/10.3389/fevo.2019.00153

SOE, K. M. *et al.* Evaluation of effective Myanmar *Bradyrhizobium* strains isolated from Myanmar soybean and effects of coinoculation with *Streptomyces griseoflavus* P4 for nitrogen fixation. **Soil Science and Plant Nutrition**, [*s. l*], v. 59, n. 3, p. 361-370, jun. 2013. http://dx.doi.org/10.1080/00380768.2013.794437

SOLANS, M. *et al.* Inoculation with native actinobacteria may improve desert plant growth and survival with potential use for restoration practices. **Microbial Ecology**, [*s. l*], v. 83, n. 2, p. 380-392, abr. 2021. http://dx.doi.org/10.1007/s00248-021-01753-4

STADIE, J. *et al.* Metabolic activity and symbiotic interactions of lactic acid bacteria and yeasts isolated from water kefir. **Food Microbiology**, [S.L.], v. 35, n. 2, p. 92-98, sep. 2013. http://dx.doi.org/10.1016/j.fm.2013.03.009

TAKAHASHI, Y.; NAKASHIMA, T.. Actinomycetes, an Inexhaustible Source of Naturally Occurring Antibiotics. **Antibiotics**, [*s. l*], v. 7, n. 2, p. 45, May. 2018. http://dx.doi.org/10.3390/antibiotics7020045

TANVIR, R. *et al.* Endophytic Actinomycetes in the Biosynthesis of Bioactive Metabolites: Chemical Diversity and the Role of Medicinal Plants. In: ATTA-UR-RAHMAN (ed.). **Studies in Natural Products Chemistry**. 60. ed. Amsterdam: Elsevier, 2019. Chap. 11. p. 399-424. Https://doi.org/10.1016/B978-0-444-64181-6.00011-5

THE SOCIETY FOR ACTINOMYCETES JAPAN. **Atlas of Actinomycetes**. Tokyo: Asakura Publishing, 1997.

UL-HAMID, A. Introduction. **A Beginners' guide to scanning electron microscopy** [S.L.], p. 1-14, 2018. Springer International Publishing. http://dx.doi.org/10.1007/978-3-319-98482-7_1

VAN BERGEIJK, D. A *et al.* Ecology and genomics of actinobacteria: new concepts for natural product discovery. **Nature Reviews Microbiology**, [*s. l*], v. 18, n. 10, p. 546-558, jun. 2020. http://dx.doi.org/10.1038/s41579-020-0379-y

VAN TATENHOVE-PEL, R. J. *et al.* Population dynamics of microbial cross-feeding are determined by co-localization probabilities and cooperation-independent cheater growth. **The Isme Journal**, [S.L.], v. 15, n. 10, p. 3050-3061, May 2021. http://dx.doi.org/10.1038/s41396-021-00986-y

VENTURA, M. *et al.* Genomics of Actinobacteria: tracing the evolutionary history of an ancient phylum. **Microbiology and Molecular Biology Reviews**, [s. l], v. 71, n. 3, p. 495-548, sep. 2007. http://dx.doi.org/10.1128/mmbr.00005-07

VERMA, V. C. *et al.* Endophytic actinomycetes from *Azadirachta indica* A. Juss.: isolation, diversity, and anti-microbial activity. **Microbial Ecology**, [S.L.], v. 57, n. 4, p. 749-756, oct. 2008. http://dx.doi.org/10.1007/s00248-008-9450-3

WHEATLEY, R. M. *et al.* Lifestyle adaptations of *Rhizobium* from rhizosphere to symbiosis. **Proceedings Of The National Academy Of Sciences**, [S.L.], v. 117, n. 38, p. 23823-23834, 8 sep. 2020. http://dx.doi.org/10.1073/pnas.2009094117

WU, Y. *et al.* Soil biofilm formation enhances microbial community diversity and metabolic activity. **Environment International**, [s.l], v. 132, p. 105116, nov. 2019. http://dx.doi.org/10.1016/j.envint.2019.105116

YANG, J. *et al.* Mechanisms underlying legume-rhizobium symbioses. **Journal of Integrative Plant Biology**, [s. l], p. 244-267, Dec. 2021. http://dx.doi.org/10.1111/jipb.13207

YIN, Q. *et al. Euzebya rosea* sp. nov., a rare actinobacterium isolated from the East China sea and analysis of two genome sequences in the genus *Euzebya*. **International Journal of Systematic and Evolutionary Microbiology**, [s. l], v. 68, n. 9, p. 2900-2905, Sep. 2018. http://dx.doi.org/10.1099/ijsem.0.002917

ZARDAK, S. G. *et al.* Effects of using arbuscular mycorrhizal fungi to alleviate drought stress on the physiological traits and essential oil yield of fennel. **Rhizosphere**, [S.L.], v. 6, p. 31-38, jun. 2018. http://dx.doi.org/10.1016/j.rhisph.2018.02.001

ZHENG, T. *et al.* Improvement of α-amylase to the metabolism adaptions of soil bacteria against PFOS exposure. **Ecotoxicology And Environmental Safety**, [S.L.], v. 208, p. 111770, jan. 2021. http://dx.doi.org/10.1016/j.ecoenv.2020.111770

ZHUANG, X. *et al.* Characterization of *Streptomyces piniterrae* sp. nov. and identification of the putative gene cluster encoding the biosynthesis of heliquinomycins. **Microorganisms**, [s. l], v. 8, n. 4, p. 495, mar. 2020. http://dx.doi.org/10.3390/microorganisms8040495

TABLE OF CONTENTS

yes I want morebooks!

Buy your books fast and straightforward online - at one of world's fastest growing online book stores! Environmentally sound due to Print-on-Demand technologies.

Buy your books online at
www.morebooks.shop

Kaufen Sie Ihre Bücher schnell und unkompliziert online – auf einer der am schnellsten wachsenden Buchhandelsplattformen weltweit! Dank Print-On-Demand umwelt- und ressourcenschonend produzi ert.

Bücher schneller online kaufen
www.morebooks.shop